Preparation for the Next Potential Pandemic—Chikungunya, Dengue, Zika and Other Viruses

Preparation for the Next Potential Pandemic—Chikungunya, Dengue, Zika and Other Viruses

Guest Editor

Zoltan Vajo

MDPI

Basel • Beijing • Wuhan • Barcelona • Belgrade • Novi Sad • Cluj • Manchester

Guest Editor
Zoltan Vajo
Department of Family Medicine
Semmelweis University
Budapest
Hungary

Editorial Office
MDPI AG
Grosspeteranlage 5
4052 Basel, Switzerland

This is a reprint of the Special Issue, published open access by the journal *Viruses* (ISSN 1999-4915), freely accessible at: https://www.mdpi.com/journal/viruses/special_issues/456F9XCARM.

For citation purposes, cite each article independently as indicated on the article page online and as indicated below:

Lastname, A.A.; Lastname, B.B. Article Title. *Journal Name* **Year**, *Volume Number*, Page Range.

ISBN 978-3-7258-7833-8 (Hbk)
ISBN 978-3-7258-7834-5 (PDF)
https://doi.org/10.3390/books978-3-7258-7834-5

Contents

About the Editor

Zoltan Vajo

Zoltan Vajo is an honorary professor of medicine, currently working at the Department of Family Medicine, Semmelweis University, Budapest, Hungary. He previously worked at the National Human Genome Institute, NIH, Bethesda, Maryland, and spent time at INSERM, Paris, France. His current area of interest is virology, immunology and mitigation processes of epidemics and pandemics. He participated in developing serological standards for pandemic influenza and creating pandemic influenza vaccines. He takes part in teaching medical students and mentoring PhD candidates.

viruses

MDPI

Editorial

Preparing for the Next Potential Pandemic—Chikungunya, Dengue, Zika and Other Viruses

Zoltan Vajo [1,*] and Csaba Laszlofy [2]

1 Department of Family Medicine, Semmelweis University Medical School, 1085 Budapest, Hungary
2 Faculty of Dentistry, University of Szeged, 6720 Szeged, Hungary
* Correspondence: zoltanvajo@gmail.com

1. Introduction

Pandemics have claimed millions of lives over millennia. The bubonic plague, smallpox, cholera, and, more recently, the Spanish flu devastated the world, resulting in more deaths than wars and enormous economic impacts. With the rapid advancement of medical knowledge during the 20th century, especially concerning infectious diseases and microbiology, it may appear that the importance of pandemics and the damage they cause will be much less concerning in the future. However, the massive threat of bird flu posed by Influenza A H5N1 and other highly pathogenic avian influenza viruses beginning in 2006, the subsequent emergence of swine flu in 2009–2010, the appearance of SARS and MERS, and, most importantly, the COVID-19 pandemic serve as wake-up calls, indicating that even with great advances in medicine, some aspects of our modern lifestyle, such as urbanization, massive sports events and concerts (with tens of thousands of people concentrating in small areas), indoor malls, and travel habits may actually increase the likelihood of future pandemics, reducing the effects of the epidemiological advances we have made.

Newly emerging or mutating variants of previously known microorganisms may continue to present potential for further pandemics. This Special Issue focuses on such viruses, including chikungunya, dengue, Zika, and others with pandemic potential.

2. Chikungunya

The Chikungunya virus is classified as an arbovirus, part of the Togaviridae family, with mosquitoes being its most important vector [1]. The most common symptoms of the infection are fever, muscle and joint pain, and rashes. Most people recover spontaneously, but in 30–40% of the affected patients, infection can lead to subacute joint inflammation, lasting weeks or more.

Thus far, at least 317,000 Chikungunya infections, with 135 Chikungunya-associated fatalities, have been confirmed across several separate geographic areas. Infections have been confirmed in all continents except Australia [2]. The ongoing appearance of Chikungunya viral infections has demonstrated that despite all measures taken so far to manage Chikungunya infections, it remains a significant concern not only in warmer climates but also in other parts of the world.

There are no specific antiviral treatment options available for chikungunya virus infection, making managing the threat posed by this virus even more challenging. Vaccines against the chikungunya virus, including live-attenuated and virus-like particle vaccines, are under development and being subjected to intense testing [3,4]. Two chikungunya virus vaccines have been licensed in the United States, namely, a live-attenuated vaccine

 https://doi.org/10.3390/v18010037

(IXCHIQ) and a virus-like particle vaccine (VIMKUNYA), with the former being currently suspended due to safety concerns, especially for elderly patients [5]. This context further highlights the challenges we face with the next potential pandemic.

3. Dengue

Dengue fever is an acute febrile illness caused by one of the four serotypes of Dengue virus. It has variable clinical manifestations, ranging from asymptomatic to severe dengue hemorrhagic fever and shock. Extended and unusual presentations involving the heart, liver, kidneys, muscles, and brain have also been reported [6]. The Dengue virus is a flavivirus with a single-stranded RNA genome and comprises three structural proteins, a capsid, an envelope, and a membrane. The global incidence of dengue has recently increased significantly, with almost half of the world's population now potentially at risk of infection [7]. Since there is no antiviral medication specifically for treating dengue fever, treatment options are severely limited, mainly constituting symptom relief and support. Infection with one serotype of the virus does not guarantee protection from the other serotypes, and reinfection with a different serotype may result in more severe symptoms [7,8]. Thus, the development of a safe tetravalent vaccine that produces a balanced immune response to all four serotypes has been a longstanding goal, and the most promising approach involves live attenuated virus vaccines, which represent their own significant challenges. Besides vaccination, effective dengue control also involves vector control, educational programs, and epidemiological surveillance [9].

4. Zika

The Zika virus encompasses a positive single-stranded RNA genome and belongs to the flaviviridae family [10]. The virus was first isolated in Uganda in 1947. It is considered an arbovirus transmitted primarily by mosquitoes. However, at least in some cases, person-to-person spread has also been verified [11,12]. Most Zika virus infections are asymptomatic or cause only mild, general symptoms such as the common cold. Frequent symptoms include a low-grade fever, malaise, myalgias, skin rashes, headaches, and joint pain, most commonly in the hands and feet. Importantly, Zika virus infections during pregnancy can cause serious complications, including microcephaly, along with Guillain–Barré syndrome in adults [13].

Despite recognition of the threat it poses, decades after its discovery, no vaccines have been granted approval against Zika virus, making the population globally vulnerable to potential future outbreaks [14]. Nonetheless, developments in several different pathways have recently been made to induce immunity against the Zika virus, and, together with the traditional inactivated and live-attenuated viral vaccines and the more recent DNA vector, mRNA-based, and recombinant protein vaccines, multiple platforms are moving towards randomized, controlled human trials.

5. Highly Pathogenic Avian Influenza (HPAI)

Highly pathogenic avian influenza viruses, mostly influenza A H5N1 and H7N9 have been considered a potential pandemic threat since the early 2000s. These strains cause severe disease with high fatality rates when transmitted from birds to humans, with a case/fatality ratio of up to 59% [15]. Human-to-human transmission has occurred, but it is infrequent [16]. The WHO set level 3/6 pandemic alertness in 2006 and recommended developing and stockpiling vaccines [17].

HPAI poses some unique challenges. Conventional diagnostic assays for detecting influenza infection and/or verifying protection, whether natural or vaccine-induced, have been found to show extreme intra- and interlaboratory variability. The traditionally used

https://doi.org/10.3390/v18010037

chicken-red-blood-cell based hemagglutination inhibition assays have been found to be insensitive and unreliable; therefore, the use of horse erythrocytes has been proposed to assess response to infection and/or vaccination [18]. Due to the information above, the World Health Organization (WHO) established H5 reference laboratories as part of the WHO Global Influenza Surveillance and Response System [19]. An additional challenge posed by HPAI is that many strains have been found to be oseltamivir-resistant [20].

HPAI strains continue to evolve and represent a threat, as recently highlighted by a fatal case in the state of Washington, which was the first confirmed global human case of a rare H5N5 strain [21].

6. Closing Thoughts

Many of the viruses that might lead to a pandemic pose several challenges. For instance, recent data suggest that Chikungunya is severely underreported and often misdiagnosed as dengue fever due to a somewhat-similar symptom profile [22].

Developing vaccines—whether the more conventional forms, such as whole-virion inactivated or attenuated, subunit, and recombinant protein, or the more novel VLP, DNA vector, and mRNA-based vaccines—is one of the most important ways of preparing for potential pandemics. Finding reliable markers of immunity, weather from infection or vaccination, poses additional challenges. This was very evident during the previous pandemics—such as those caused by Influenza A(H1N1)pdm09 and, most recently, by SARS-CoV-2—when reliable correlates of protection were difficult to standardize.

Furthermore, developing novel antivirals and repurposing previously well-known medications are also essential parts of strategies for addressing potential pandemic viruses. This was made very clear during the COVID-19 pandemic, where both methods were successfully used. Discussing public health, epidemiological, pharmaceutical, immunological, and other measures that might help us avoid or at least mitigate future pandemics and thus prevent loss of life and decrease economic burdens is of utmost importance.

Conflicts of Interest: The authors declare no conflict of interest.

References

1. Martelossi-Cebinelli, G.; Carneiro, J.A.; Yaekashi, K.M.; Bertozzi, M.M.; Bianchini, B.H.S.; Rasquel-Oliveira, F.S.; Zanluca, C.; Duarte Dos Santos, C.N.; Arredondo, R.; Blackburn, T.A.; et al. A Review of the Biology of Chikungunya Virus Highlighting the Development of Current Novel Therapeutic and Prevention Approaches. *Pathogens* **2025**, *14*, 1047. [CrossRef] [PubMed]
2. European Centre for Disease Prevention and Control. Chikungunya Virus Disease Worldwide Overview. Situation Update. August 2025. Available online: https://www.ecdc.europa.eu/en/chikungunya-monthly (accessed on 31 October 2025).
3. Simone, B.; Lienert, F. Post-authorisation experience and reported adverse events following use of a virus-like particle chikungunya vaccine, United States and Germany, up to August 2025. *Eurosurveillance* **2025**, *30*, 2500792. [CrossRef] [PubMed] [PubMed Central]
4. Maure, C.; Khazhidinov, K.; Kang, H.; Auzenbergs, M.; Moyersoen, P.; Abbas, K.; Santos, G.M.L.; Medina, L.M.H.; Wartel, T.A.; Kim, J.H.; et al. Chikungunya vaccine development, challenges, and pathway toward public health impact. *Vaccine* **2024**, *42*, 126483. [CrossRef] [PubMed]
5. US Food and Drug Administration. FDA Update on the Safety of Ixchiq (Chikungunya Vaccine, Live). Available online: https://www.fda.gov/vaccines-blood-biologics/safety-availability-biologics/fda-update-safety-ixchiq-chikungunya-vaccine-live#:~:text=The%20FDA%E2%80%99s%20Center%20for%20Biologics%20Evaluation%20and,public%20would%20pose%20a%20danger%20to%20health (accessed on 9 November 2025).
6. Tsheten, T.; Clements, A.C.A.; Gray, D.J.; Adhikary, R.K.; Furuya-Kanamori, L.; Wangdi, K. Clinical predictors of severe dengue: A systematic review and meta-analysis. *Infect. Dis. Poverty* **2021**, *10*, 123. [CrossRef] [PubMed] [PubMed Central]
7. Kularatne, S.A.; Dalugama, C. Dengue infection: Global importance, immunopathology and management. *Clin. Med.* **2022**, *22*, 9–13. [CrossRef] [PubMed] [PubMed Central]

https://doi.org/10.3390/v18010037

8. Andrade, E.H.; Figueiredo, L.B.; Vilela, A.P.; Rosa, J.C.; Oliveira, J.G.; Zibaoui, H.M.; Araújo, V.E.; Miranda, D.P.; Ferreira, P.C.; Abrahão, J.S.; et al. Spatial-Temporal Co-Circulation of Dengue Virus 1, 2, 3, and 4 Associated with Coinfection Cases in a Hyperendemic Area of Brazil: A 4-Week Survey. *Am. J. Trop. Med. Hyg.* **2016**, *94*, 1080–1084. [CrossRef] [PubMed] [PubMed Central]
9. Khetarpal, N.; Khanna, I. Dengue Fever: Causes, Complications, and Vaccine Strategies. *J. Immunol. Res.* **2016**, *2016*, 6803098. [CrossRef] [PubMed] [PubMed Central]
10. Faye, O.; Freire, C.C.; Iamarino, A.; Faye, O.; de Oliveira, J.V.; Diallo, M.; Zanotto, P.M.; Sall, A.A. Molecular evolution of Zika virus during its emergence in the 20(th) century. *PLoS Neglected Trop. Dis.* **2014**, *8*, e2636. [CrossRef] [PubMed]
11. Lazear, H.M.; Diamond, M.S. Zika Virus: New Clinical Syndromes and Its Emergence in the Western Hemisphere. *J. Virol.* **2016**, *90*, 4864–4875. [CrossRef] [PubMed] [PubMed Central]
12. Baud, D.; Musso, D.; Vouga, M.; Alves, M.P.; Vulliemoz, N. Zika virus: A new threat to human reproduction. *Am. J. Reprod. Immunol.* **2017**, *77*, e12614. [CrossRef] [PubMed]
13. Cauchemez, S.; Besnard, M.; Bompard, P.; Dub, T.; Guillemette-Artur, P.; Eyrolle-Guignot, D.; Salje, H.; Van Kerkhove, M.D.; Abadie, V.; Garel, C.; et al. Association between Zika virus and microcephaly in French Polynesia, 2013–2015: A retrospective study. *Lancet* **2016**, *387*, 2125–2132. [CrossRef] [PubMed]
14. Lee, H.J.; Abu Bakar, S.; Shin, O.S. An updated review of Zika virus vaccine development. *Clin. Exp. Vaccine Res.* **2025**, *14*, 325–334. [CrossRef] [PubMed]
15. Yang, Y.; Halloran, M.E.; Sugimoto, J.D.; Longini, I.M., Jr. Detecting human-to-human transmission of avian influenza A (H5N1). *Emerg. Infect. Dis.* **2007**, *13*, 1348–1353. [CrossRef] [PubMed] [PubMed Central]
16. Ungchusak, K.; Auewarakul, P.; Dowell, S.F.; Kitphati, R.; Auwanit, W.; Puthavathana, P.; Uiprasertkul, M.; Boonnak, K.; Pittayawonganon, C.; Cox, N.J.; et al. Probable person-to-person transmission of avian influenza A (H5N1). *N. Engl. J. Med.* **2005**, *352*, 333–340. [CrossRef] [PubMed]
17. Dennis, C. Flu-vaccine makers toil to boost supply. *Nature* **2006**, *440*, 1099. [CrossRef] [PubMed]
18. Stephenson, I.; Wood, J.M.; Nicholson, K.G.; Charlett, A.; Zambon, M.C. Detection of anti-H5 responses in human sera by HI using horse erythrocytes following MF59-adjuvanted influenza A/Duck/Singapore/97 vaccine. *Virus Res.* **2004**, *103*, 91–95. [CrossRef] [PubMed]
19. WHO H5 Reference Laboratories and the Terms of Reference. Available online: https://www.who.int/initiatives/global-influenza-surveillance-and-response-system/h5-reference-laboratories (accessed on 17 December 2025).
20. de Jong, M.D.; Tran, T.T.; Truong, H.K.; Vo, M.H.; Smith, G.J.; Nguyen, V.C.; Bach, V.C.; Phan, T.Q.; Do, Q.H.; Guan, Y.; et al. Oseltamivir resistance during treatment of influenza A (H5N1) infection. *N. Engl. J. Med.* **2005**, *353*, 2667–2672. [CrossRef] [PubMed]
21. Washington State Department of Health. Grays Harbor County Resident Dies from Complications of Avian Influenza. Available online: https://doh.wa.gov/newsroom/grays-harbor-county-resident-dies-complications-avian-influenza (accessed on 17 December 2025).
22. Ribas Freitas, A.R.; Pinheiro Chagas, A.A.; Siqueira, A.M.; Pamplona de Góes Cavalcanti, L. How much of the current serious arbovirus epidemic in Brazil is dengue and how much is chikungunya? *Lancet Reg. Health Am.* **2024**, *34*, 100753. [CrossRef] [PubMed] [PubMed Central]

https://doi.org/10.3390/v18010037

viruses

MDPI

Article

Long-Term Stabilization of Dengue Virus RNA at 37 °C for 14 Months Using Silk Fibroin Films

Nuo Wang [1,†], Ping He [2,†], Bohan Xu [1,3,†], Hongping Wei [1,3,*] and Junping Yu [1,3,*]

1 State Key Laboratory of Virology and Biosafety, Wuhan Institute of Virology, Chinese Academy of Sciences, Wuhan 430071, China; wangnuo@wh.iov.cn (N.W.); xubohan22@mails.ucas.ac.cn (B.X.)

2 Chongqing Key Laboratory of Highly Pathogenic Microbes, Chongqing Disease Prevention and Public Health Research Center Construction Program, Chongqing Center for Disease Control and Prevention (Chongqing Academy of Preventive Medicine), Chongging 400707, China; peace192@163.com

3 University of Chinese Academy of Sciences, Beijing 100049, China

* Correspondence: hpwei@wh.iov.cn (H.W.); yujp@wh.iov.cn (J.Y.); Tel.: +86-27-87998873 (H.W.); +86-27-87998263 (J.Y.)

† These authors contributed equally to this work.

Abstract: Diagnosis of dengue virus infections typically relies on RT-PCR-based methods, for which reliable positive controls are essential. Viral RNA is an ideal positive control, but its inherent instability poses a major challenge. Herein, we report a simple and effective method for stabilizing dengue virus RNA by immobilizing it onto silk fibroin films (RNA-SFFs). We evaluated various substrate surfaces for RNA-SFFs preparation and found that the inner surface of sealable bags is optimal for uniform film formation and easy harvesting. Screening different silk fibroin concentrations revealed that even low concentrations (2.8%) effectively preserved RNA well and kept Ct constant for up to 16 days at 25 °C, 37 °C, and even 45 °C (extreme weather for transportations). Due to its rapid film formation and ease of peeling, 7% silk fibroin was selected. Notably, the RNA-SFFs demonstrated robust resistance to UV irradiation, with no significant Ct value changes after 4 h of exposure. Long-term stability testing at −20 °C, 25 °C, and 37 °C showed that dengue serotype 1–4 RNA-SFFs remained stable for the entire duration of the study—up to 56 weeks (approximately 14 months)—at all tested temperatures. These results demonstrate that RNA-SFFs are highly stable, portable, and practical as positive controls for dengue diagnostics, with strong potential for use in on-site and resource-limited settings.

Keywords: positive controls; RNA; dengue virus; silk fibroin film; stability

1. Introduction

Dengue, an arboviral disease caused by dengue viruses and transmitted by *Aedes* mosquitoes, can lead to mild or severe sickness. The symptoms include fever, aches (such as eye pain, typically behind the eyes, muscle, joint, or bone pain), nausea or rashes. Approximately 5% of individuals who develop dengue will progress to severe dengue, which may result in shock, internal bleeding, and even death [1,2]. The dengue virus comprises four distinct serotypes (DENV-1 to DENV-4), contributing to its complex epidemiology. In 2024, the World Health Organization's (WHO) reported a record 14.4 million dengue cases globally (https://worldhealthorg.shinyapps.io/dengue_global/, accessed on 15 October 2025), which is more than double the 7 million cases recorded in 2023 [3]. The real situation might be more serious. One modeling estimate indicates that dengue fever affects more than 400 million people annually worldwide and causes around 22,000 deaths across over 100 countries. Given the lack of specific antiviral treatment, early diagnosis continues to play a critical role in enabling timely treatment and minimizing the risk of severe dengue [2,4].

Reverse transcription-polymerase chain reaction (RT-PCR) remains the benchmark method for the early and accurate detection of dengue virus infection [5,6]. To ensure the reliability and validity of RT-PCR, the inclusion of positive controls or reference materials in each run is essential. Genomic RNA has become a widely adopted reference standard in molecular diagnostics and assay calibration due to its full-length genomic sequence, ease of production, and suitability for scalable manufacturing [7,8]. However, RNA molecules are inherently unstable and highly susceptible to enzymatic (e.g., ribonuclease-mediated) and chemical degradation (e.g., hydrolysis, oxidation), necessitating immediate stabilization via storage at ultra-low temperatures or under desiccated conditions [7,9–11]. This requirement presents significant challenges for resource-limited settings.

To circumvent these constraints, various ambient-temperature stabilization strategies have been investigated. Lyophilization (freeze-drying) [12], encapsulation within metallic capsules [10,11], or entrapment in silica-based microparticles [13] offer enhanced stability but are often prohibitively expensive, labor-intensive, or ill-suited for deployment in resource-limited or field settings.

Silk fibroin is a natural protein polymer derived from the cocoons of the domesticated silkworm *Bombyx mori*. In its native state, silk fibroin adopts a semicrystalline architecture comprising approximately 65% crystalline domains and 35% amorphous regions, which synergistically contribute to the mechanical resilience and structural robustness of the cocoons. Silk fibroin exhibits high solubility in water and can be processed into a regenerated material with a water-stable structure, featuring a predominant β-sheet crystalline conformation. This conformation endows the protein-based matrix with exceptional thermal stability, mechanical strength under tension and resistance to chemical degradation [14,15]. Owing to these unique physicochemical properties—particularly its biocompatibility and ability to stabilize labile biomolecules—silk fibroin has emerged as a highly promising matrix for the encapsulation and preservation of thermosensitive biological agents, including whole blood, DNA, and RNA [14,16–21].

In this study, silk fibroin is employed to form films incorporating dengue viral RNA, serving as reference materials to support the development and evaluation of RT-PCR-based dengue virus diagnostic assays. The resulting RNA-silk fibroin films (RNA-SFFs) effectively stabilize dengue viral RNA, with no significant change in RT-qPCR Ct values observed after storage for at least 14 months at 37 °C (longer durations have not yet been evaluated). RNA-SFFs prepared from various concentrations of silk fibroin maintain RNA stability at 25 °C, 37 °C and 45 °C for 16 days. Additionally, the RNA-SFFs exhibit resistance to UV radiation, as evidenced by stable Ct values following 4 h of exposure to UV light at an intensity of over 90 $\mu W/cm^2$. These results suggest minimal RNA degradation under thermal and environmental stress. Therefore, the developed RNA-SFFs represent a promising cold-chain-free alternative for the long-term storage and transport of reference materials, with potential utility in resource-limited and field-deployable diagnostic settings.

2. Materials and Methods

2.1. Materials

Dengue virus serotypes 1 through 4 were obtained from clinical specimens derived from infected individuals and propagated in *Aedes albopictus* C6/36 cells (IVCAS9.087 in National Virus Resource Center, Wuhan Institute of Virology, CAS, Wuhan, China). C6/36 cells were cultured in T75 flasks using MEM supplemented with 10% fetal bovine serum (FBS) and 1% penicillin-streptomycin. When cells reached 70–80% confluence, they were infected with DENV-1 to DENV-4 at a multiplicity of infection (MOI) ranging from

0.01 to 0.1. The infected cultures were incubated at 37 °C in a 5% CO_2 humidified incubator until pronounced cytopathic effect (CPE) was observed (3–4 days). Viral supernatants were then collected for RNA extraction. Although the initial infectious virus titer was not determined by plaque assay or $TCID_{50}$, the viral load in the supernatant was assessed by RT-qPCR, yielding Ct values between 10 and 15, indicating high levels of viral RNA. Viral nucleic acids were isolated utilizing the QIAamp® Viral RNA Mini Kit (Qiagen, Hilden, Germany), following the manufacturer's protocol. The HiScript® II One Step qRT-PCR Probe Kit (Vazyme Biotech Co., Ltd., Nanjing, China) was used for RT-qPCR. Cocoons were purchased from Crystalgen Ningbo Biotech Ltd. (Ningbo, China). Lithium bromide and sodium carbonate were bought from Shanghai Aladdin Biochemical Technology Co., Ltd. (Shanghai, China) and Sinopharm Chemical Reagent Co., Ltd. (Shanghai, China). All primers and fluorescently labeled probes were commercially synthesized by Sangon Biotech Co., Ltd. (Shanghai, China). Nuclease-free water, essential for molecular assays to prevent degradation of nucleic acids, was sourced from Source Leaf Biotech Co., Ltd. (Shanghai, China). The primer/probe sets for the four dengue virus serotypes 1 through 4 are listed in Table 1. Ultrapure water with a resistivity of 18.2 MΩ·cm was employed in all experiments.

Table 1. Primers and probes targeting the four dengue virus serotypes for RT-qPCR.

Primer or Probe	Sequence (5′-3′)
DENV-1-F	TGTGCATTGAAGCTAAAATATCA
DENV-1-R	CGTCTTGTTCTTCCACCA
DENV-1-P	FAM-ACCACCACCGACTCAAGATGTCCAA-BHQ1
DENV-2-F	CGAGAAATACGCCTTTCAATA
DENV-2-R	CAGCATTCCAAGTGAGAATC
DENV-2-P	FAM-AACCGCGTGTCAACTGTGCAAC-BHQ1
DENV-3-F	CAACCAACGGAAGAAGAC
DENV-3-R	CGCCAACTGTGATCCAGT
DENV-3-P	FAM-AAACCGTCTATCAATATGCTGAAACGC-BHQ1
DENV-4-F	GGTTGGTGAAGAGATTCTCA
DENV-4-R	GTGGGATGGAAAGGACTC
DENV-4-P	FAM-AGCACCATCCGTAAGGGTCCT-BHQ1

2.2. Preparation and Quantification of Silk Fibroin Solution from Silk Cocoons

Silk fibroin solution was prepared following the established protocol described by Rockwood D. et al. [16], with minor modifications. Approximately 2.5 g of *Bombyx mori* silk cocoons were manually cut into fragments of nail-clipping size using sterilized scissors. To facilitate degumming, 2.12 g of sodium carbonate was dissolved in 1 L of ultrapure water, which was heated to boiling to dissolve. The cocoon pieces were introduced into the boiling sodium carbonate solution and stirred vigorously at 110–200 rpm for 50 min to remove sericin. Following cooling, the degummed silk fibers were retrieved, transferred to 1 L of fresh ultrapure water, and subjected to magnetic stirring for 20 min to remove residual salts. This washing step was repeated three times with complete water replacement. The purified fibers were then dried overnight in a fume hood. The dried silk fibroin was weighed (about 1.70 g). For dissolution, 8.1 g of lithium bromide (LiBr) was dissolved in 10 mL of ultrapure water. The dried fibers were solubilized in the LiBr solution at a ratio of 1:4 (for example, 1.7 g silk fibers to 6.8 mL LiBr solution), followed by incubation at 60 °C for 4 h to ensure complete dissolution of the protein matrix. The resulting solution was dialyzed against 1 L of ultrapure water at 4 °C for 72 h, with the dialysate replaced three times daily to eliminate residual LiBr. After dialysis, the aqueous silk fibroin solution was centrifuged at 9000× *g* and 4 °C for 20 min to remove insoluble aggregates. The supernatant was collected and subjected to a second centrifugation under identical conditions to ensure maximal clarity. The purified silk fibroin solution was stored at 4 °C for up to one month and used in subsequent experiments as required. The concentration of the prepared silk fibroin solution was determined as follows. A clean Petri dish lid was pre-weighed (mass recorded as m_1 in grams). Exactly 500 μL of the silk fibroin solution was pipetted onto the lid and dried at 60 °C for 1 h to constant weight. The lid with dried fibroin was weighed (mass recorded as m_2 in grams, and the mass difference in mass ($m_2 - m_1$). The concentration of the prepared silk fibroin solution is calculated as:

$$c_{\text{silk fibroin}} = \frac{m_2 - m_1}{0.5}$$

The unit of the protein concentration is in % *w*/*v*. For example, if the mass difference is 0.035 g, then the concentration is 0.035 divided by 0.5, which is 7.0% *w*/*v*.

2.3. Evaluation of Film Formation Efficiency of the Silk Fibroin on Various Substrate Surfaces

The prepared silk fibroin solution in Section 2.2 was quantified and adjusted to a final concentration of 7.0% *w*/*v*. To investigate the influence of substrate properties on film formation and ease of harvest, 10 μL of RNase-free water was mixed with 10 μL of the 7.0% *w*/*v* silk fibroin solution. The resulting blend was cast onto various surfaces to form circular films with a diameter of approximately 1 cm, including a Petri dish (plastic), a Petri dish (glass), aluminum foil, the inner surface of an incised new sealable plastic bag, and the interior of microcentrifuge tubes. Samples were air-dried under ambient conditions to form thin solid films. The time required for complete film formation and the ease of film detachment were assessed for each substrate.

2.4. RT-qPCR System and Protocol

The RT-qPCR assay was carried out in a total reaction volume of 20 μL, comprising 5 μL of template RNA, 10 μL of 2× one-step reaction mix, 1 μL of enzyme mix, 0.4 μL each of 10 μM forward and reverse primers, 0.4 μL of 10 μM fluorescent probe, and nuclease-free water to achieve the final volume. The RT-qPCR reactions were run in duplicate. Amplification and fluorescence detection were performed using the CFX96 Real-Time PCR Detection System (Bio-Rad, Hercules, CA, USA) or MA-6000 (Yarui Biotech, Suzhou, China) under the following thermal cycling conditions: an initial reverse transcription phase at 50 °C for 15 min, followed by initial denaturation at 95 °C for 30 s. This was succeeded by 45 consecutive cycles of amplification, each consisting of a denaturation step at 95 °C for 10 s and an annealing/extension step at 60 °C for 30 s, with fluorescence acquisition occurring during the latter phase of each cycle. The Ct thresholds for all qPCR runs were set automatically by the instrument's software (Bio-Rad CFX Manager version 3.1, Hercules, CA, USA) or MA-6000 (Yarui Biotech, Suzhou, China).

2.5. The Effect of Silk Fibroin Films (SFFs) on Ct Values of DENV-1

Evaluation of the effect of silk fibroin on Ct values of DENV-1 RNA through RT-qPCR: To assess potential matrix-induced effects of silk fibroin on RT-qPCR quantification, a dilution series of silk fibroin solutions was prepared from 7.0% *w*/*v* using a two-fold serial dilution method, yielding concentrations of 7.0%, 3.5%, 1.75%, 0.88%, and 0.44% *w*/*v*. Here, a wide range of the silk fibroin concentrations were used to identify the threshold at which silk fibroin might begin to interfere with RT-qPCR. For each concentration, 10 μL of DENV-1 RNA was mixed with 10 μL of the respective silk fibroin solution, resulting in a 1:1 (*v*/*v*) RNA-silk matrix. The mixtures were then directly subjected to RT-qPCR analysis without prior processing.

Assessment of DENV-1 RNA stability in the form of RNA-silk fibroin films (RNA-SFFs) prepared from silk fibroin solutions of varying concentrations: Silk fibroin solutions at concentrations of 7.0%, 5.6%, 4.2% and 2.8% w/v were prepared by serial dilution of the stock solution with nuclease-free water (Source Leaf Biotech Co., Ltd., Shanghai, China), which are different from the concentrations used above due to that very low concentrations may compromise film integrity. The mixtures of 10 μL of DENV-1 viral RNA and 10 μL of each silk fibroin formulation were spread onto the inner surfaces of sealed polyethylene bags and air-dried under ambient conditions for 30 min to 1 h to yield thin, solid composite films. The resulting RNA-SFFs were transferred into 1.5 mL microcentrifuge tubes and stored under controlled thermal conditions at 25 °C, 37 °C and 45 °C. The preparation process of dengue virus RNA–silk fibroin films (SFFs) is illustrated in Figure 1. At days 1, 3, 6, 9, 13 and 16, individual samples were retrieved, and each film was rehydrated by adding 500 μL of RNase-free water. Samples were vortexed vigorously to ensure complete dissolution and homogeneous resuspension of RNA. The recovered RNA was then immediately subjected to RT-qPCR, and the cycle threshold (Ct) values were recorded.

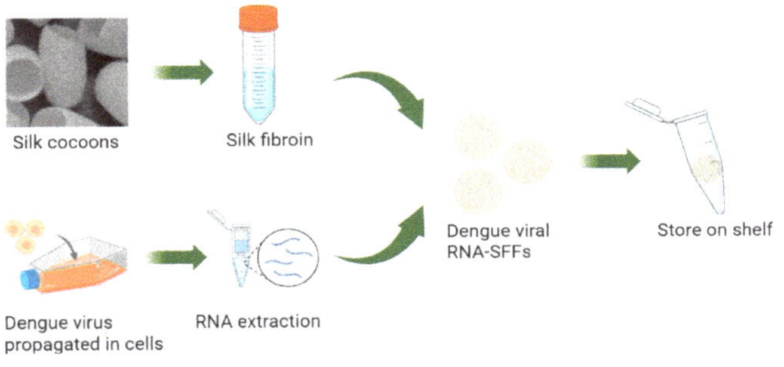

Figure 1. Experimental strategy for the preparation of dengue virus RNA–SFFs. Created in BioRender. Yu, J. (2025) https://app.biorender.com/illustrations/68d0ad9ed6333e7d2837d168?slideId=8df1f916-9599-42dd-8075-1a73ac3418d5 accessed on 29 October 2025.

2.6. Assessment of DENV-1 RNA Stability in the Form of RNA-SFFs

To evaluate the protective effect of RNA in the form of RNA-SFFs against UV-induced degradation, samples were exposed to UV radiation at an irradiance intensity exceeding 90 μW/cm^2 for durations of 1 h, 2 h, and 4 h. The irradiance level was chosen to reflect typical conditions used for surface decontamination in biological safety cabinets and clean benches. DENV-1 RNA-SFFs were prepared following the protocol outlined in Section 2.5, and naked DENV-1 RNA in solution served as the control.

2.7. Long-Term Storage of the Prepared DENV-1–4 RNA-SFFs at −20 °C, 25 °C and 37 °C

Dengue virus serotypes 1–4 RNA were individually encapsulated into SFFs using the protocol described above. The prepared RNA-SFFs were stored at −20 °C, 25 °C, and 37 °C to assess long-term RNA stability under varying thermal conditions. No special precautions against light exposure were taken during storage, and temperature fluctuations were kept to a minimum to ensure consistent experimental conditions. At 1, 2, 3, 4, 6, 8, 12, 24, 48 and 56 weeks, individual films were removed from storage and dissolved in RNase-free water according to the rehydration procedure mentioned above. The dissolved RNA was then subjected to serotype-specific RT-qPCR using corresponding primer-probe sets as listed in Table 1. Ct values were recorded at each time point to monitor Dengue RNA stability.

2.8. Data Statistics

Data are presented as the mean ± standard deviation (SD) derived from three independent replicates. Statistical significance between experimental groups at different temperature was assessed by one-way ANOVA (and nonparametric or mixed) with Dunn's multiple comparisons test using GraphPad Prism 9.0.0 (GraphPad Software Inc., San Diego, CA, USA). A probability value of $p < 0.05$ was designated as statistically significant.

3. Results

3.1. Selection of Appropriate Substrate Surfaces to Form SFFs

Initially, the film formation efficiency of silk fibroin was evaluated across various substrate surfaces. The results are summarized in Table 2. To optimize the preparation process of SFFs, an ideal substrate was sought that would enable the formation of intact, easily detachable films with a short air-drying time. As shown in Table 2, the inner surface of an incised new sealable plastic bag exhibited favorable characteristics: rapid air-drying, formation of structurally uniform films, and facile delamination without fragmentation. Given these advantages, this substrate was adopted for the preparation of RNA-loaded silk fibroin films (RNA-SFFs), where the RNA is embedded within the film matrix during casting and dried into a stable, peelable sheet for subsequent rehydration and use in RT-qPCR.

Table 2. Evaluation of the film formation efficiency of silk fibroin on various substrate surfaces.

Substrate	Air-Drying Time/min	Film Morphology	Peelability
Petri dish (polystyrene)	60	Brittle and prone to cracking	Difficult to peel intact; required multiple attempts with residual adhesion
Petri dish (glass)	40	Powder-like morphology, no continuous film structure formed	Not peelable; only removable via scraping
Aluminum foil	40	Adhered strongly to the surface; no visible intact film	Unable to be peeled off without damaging or fragmenting
Sealable plastic bag	40	Uniform, flexible and structurally intact film	Easily removable without fragmentation
Centrifuge tube	>60	Mixture remained liquid after more than one hour, no drying or film formation observed	Not applicable

3.2. Assessment of the Effect of Silk Fibroin at Varying Concentrations on DENV-1 Genomic RNA Detection

We first evaluated the effect of silk fibroin on Ct values of DENV-1 viral RNA by RT-qPCR. The results are shown in Figure 2. The figure displayed no significant differences between DENV-1 viral RNA alone (the control group is DENV-1 RNA with 0% silk fibroin) and samples containing silk fibroin at various concentrations of the silk fibroin, even at 7% *w*/*v* (the corresponding final concentration of silk fibroin in the RNA samples was 3.5%). This indicates that silk fibroin had no obvious effect on the Ct value of DENV-1 viral RNA.

3.3. Evaluation of the Stability of DENV-1 RNA in SFFs Prepared with Silk Fibroin at Varying Concentrations

DENV-1 RNA was mixed with silk fibroin solutions of four silk fibroin concentrations (7.0%, 5.6%, 4.2%, and 2.8% *w*/*v*) and dried to form RNA-silk fibroin films (RNA-SFFs) to evaluate RNA stability in the SFF format under different storage conditions at 25 °C, 37 °C and 45 °C. The results are presented in Figure 3. The data, along with statistical

p values and z values, are provided in the Supplementary Materials. A p value > 0.05 indicates no significant difference, while the z value reflects the magnitude of intergroup differences. The results indicated that at all tested concentrations, RT-qPCR Ct values remained stable throughout the 16-day storage period, even under the most thermally stressful condition (45 °C), indicating that silk fibroin effectively preserves RNA well in the SFF format for at least 16 days at elevated temperatures. Notably, RNA-SFFs prepared with 7.0% w/v silk fibroin exhibited the narrowest confidence intervals and the most consistent Ct values across all time points and temperatures, as evidenced by the smallest error bars in Figure 3D (with the smallest fluctuations for all the temperatures with 7.0% w/v silk fibroin (7.0%: 29.4 ± 0.45; 5.6%: 29.7 ± 0.64; 4.2%: 30.0 ± 0.75; 2.8%: 30.2 ± 0.61). This enhanced reproducibility is likely attributable to the formation of denser, more cohesive films at the higher silk fibroin concentration, which facilitated complete and consistent delamination from the substrate surface, thereby improving sample recovery uniformity and experimental repeatability. Based on these results, a silk fibroin concentration of 7.0% w/v was selected for subsequent experiments to ensure optimal film quality and assay consistency.

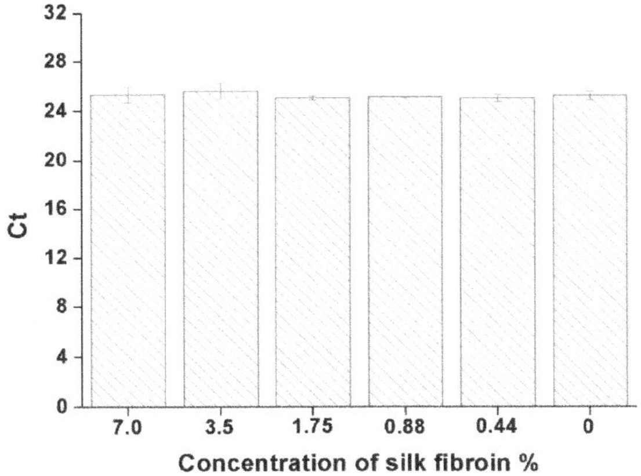

Figure 2. Impact of silk fibroin concentration on the RT-qPCR Ct values of DENV-1 RNA. Data are presented as the mean Ct value from three independent experiments, with error bars indicating the standard deviation (SD). Each concentration of silk fibroin (7.0%, 3.5%, 1.75%, 0.88%, and 0.44% w/v) was tested in a 1:1 (v/v) mixture with DENV-1 RNA. The control group is DENV-1 RNA with 0% silk fibroin. No significant variation in Ct values was observed across the concentration range. Statistical significance was determined by ANOVA vs. 0% silk fibroin (p values and z values, which reflect the magnitude of the group difference, are listed in the Supplementary Materials). No significance was observed for all the tested silk fibroin concentrations (p > 0.05).

3.4. UV Resistance of DENV-1 RNA-SFFs

DENV-1 RNA-SFFs, prepared according to the aforementioned protocol, were exposed to UV radiation at an irradiance intensity exceeding 90 μW/cm^2, which was selected based on the typical irradiance level used for surface decontamination in biological safety cabinets and clean benches, for 1 h, 2 h and 4 h to evaluate the protective capacity of DENV-1 RNA-SFFs against UV-induced RNA degradation. In contrast to the RNA-SFFs, naked DENV-1 RNA in solution exhibited a time-dependent increase in RT-qPCR Ct values with prolonged UV exposure (Figure 4), indicating progressive RNA fragmentation and reduced amplifiability. By contrast, RNA within the silk fibroin films maintained stable Ct values across all time points, with no statistically significant differences observed after 1, 2, or 4 h of irradiation ($p > 0.05$). These results demonstrate that the silk fibroin films effectively shields RNA from UV-induced photodegradation. The RNA molecules are preserved well within the film, likely due to the physical barrier and radical-scavenging properties of the silk protein network. Thus, silk fibroin confers robust protection against environmental UV stress, highlighting its potential as a stabilizing biomaterial for RNA preservation under challenging conditions.

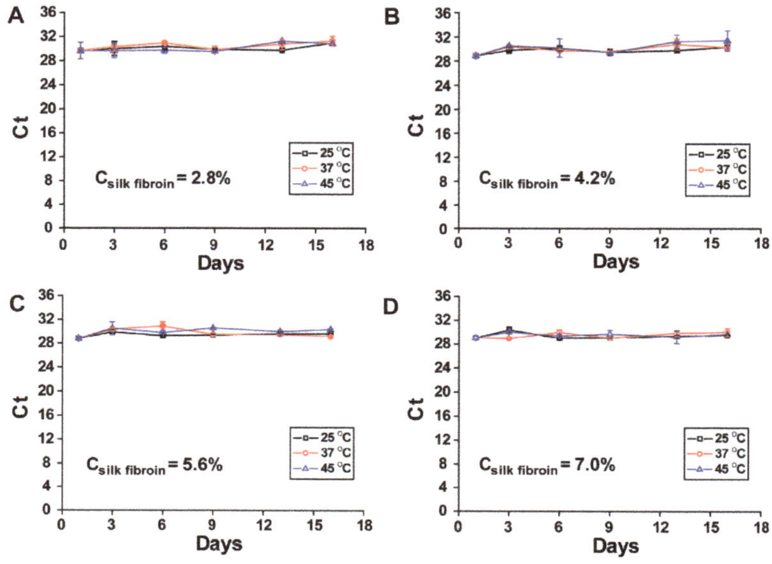

Figure 3. Stability of DENV-1 RNA in RNA-SFFs prepared with silk fibroin at varying concentrations. (**A**). 2.8% *w/v*; (**B**). 4.2% *w/v*; (**C**). 5.6% *w/v*; (**D**). 7.0% *w/v*. Each panel shows RT-qPCR Ct values of RNA in SFFs over time at 25 °C, 37 °C and 45 °C, indicating minimal degradation of RNA across all concentrations. The highest consistency was observed at 7.0% *w/v*. All data are presented as mean ± SD of three independent experiments.

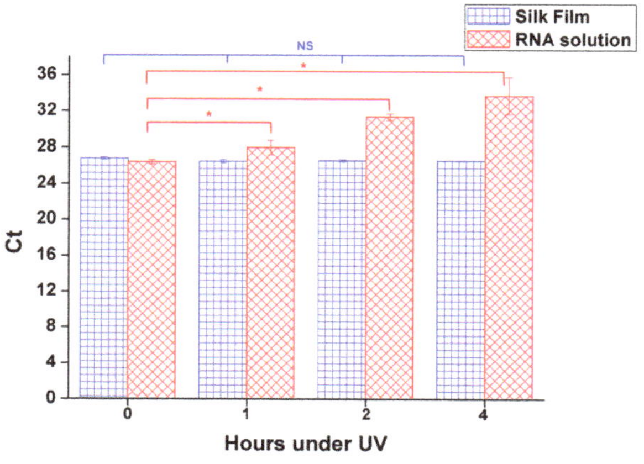

Figure 4. Stability of DENV-1 in RNA-SFFs and free RNA under UV irradiation at increasing exposure durations (1 h, 2 h and 4 h). Each data point represents the mean Ct value from three independent experiments, and error bars indicate SD. Statistical significance was determined by ANOVA vs. 0 h of UV irradiation. p and z values are displayed in the Supplementary Materials. * $p < 0.05$ for free RNA (indicating degradation); NS for RNA-SFFs (no significant change), showing effective stabilization.

3.5. Long-Term Stability of Dengue Serotypes 1–4 RNA in RNA-SFFs Under Different Temperature Conditions

To evaluate the long-term stability of dengue viral RNA in RNA-SFFs, DENV-1–4 RNA-SFFs were stored under three distinct temperature conditions: −20 °C, 25 °C and 37 °C, for an extended period of 56 weeks (approximately 14 months). All data, including statistical p and z values, are available in the Supplementary Materials. Plots of Ct values vs. weeks are presented in Figure 5. The mean Ct values for the first six time points in Figure 5 are from three independent experiments, while the data for the last four time points are from two independent experiments. As shown, DENV-2 RNA (Figure 5B) exhibited a lower Ct value at the one-week time point compared to subsequent measurements. This may reflect an initial measurement variability or a transient shift in Ct values between week one and week two, after which the values stabilized and remained consistent for the remainder of the study. No significant differences were observed in Ct values between any testing time point and the reference time point (week one, or week two for DENV-2); $p > 0.05$, see Supplementary Materials (Data for Figure 5). Notably, all four dengue serotypes demonstrated excellent stability over the 56-week duration, with minimal fluctuation in Ct values across all storage temperatures. Importantly, statistical analysis revealed no significant differences in Ct values between samples stored at −20 °C and those maintained at 25 °C or 37 °C at any tested time points ($p > 0.05$). The p values are listed in the Supplementary Materials (Data for Figure 5). This indicates that RNA-SFFs effectively preserve RNA even under elevated, non-frozen conditions. Collectively, these findings demonstrate that RNA-SFFs provide robust protection for multivalent dengue viral RNA over prolonged periods, maintaining molecular stability for up to 56 weeks without the need for cold-chain storage. These results highlight the potential of SFFs as a versatile and reliable platform for ambient-temperature biostabilization of RNA-based diagnostics.

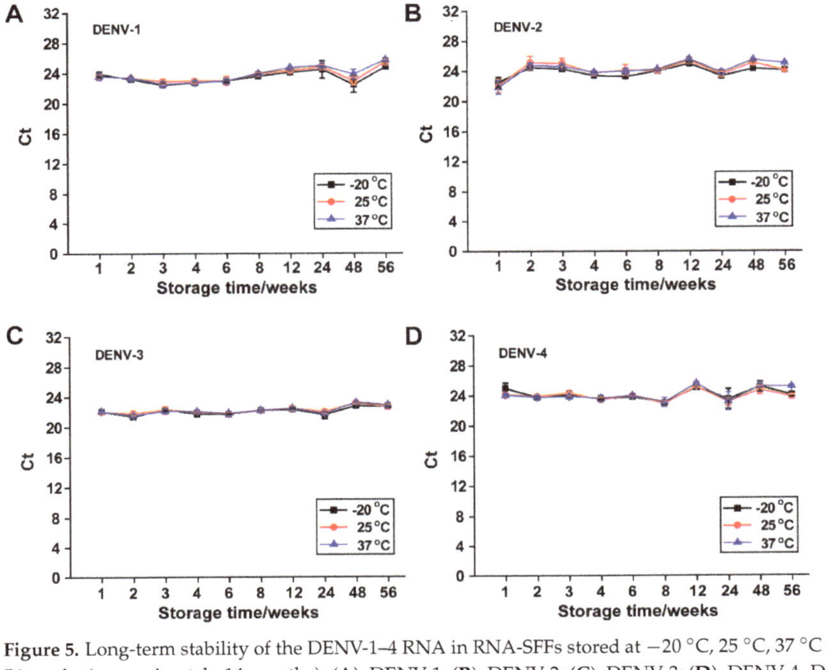

Figure 5. Long-term stability of the DENV-1–4 RNA in RNA-SFFs stored at −20 °C, 25 °C, 37 °C for 56 weeks (approximately 14 months). (**A**). DENV-1; (**B**). DENV-2; (**C**). DENV-3; (**D**). DENV-4. Data shown are means ± SD.

4. Discussion

Positive controls play a pivotal role in RT-PCR-based assays for the diagnosis of viral infections. Dengue viral genomic RNA serves as a critical positive control with advantages of containing all targets of dengue viruses to validate all RT-PCR-based diagnostic kits. However, RNA is inherently labile and highly susceptible to degradation by RNases, chemicals and environmental stressors such as heat and UV radiation. The storage and transport of RNA-based controls depend on cold-chain infrastructure, which limits the practical utility in resource-limited or field-deployable diagnostic settings.

In this study, we demonstrate that dengue RNA-SFFs can effectively stabilize dengue genomic RNA under challenging conditions. Specifically, RNA in SFFs remains stable for up to 14 months at 37 °C, up to 16 days at 45 °C, and for a minimum of 4 h under intense UV irradiation (>90 $\mu W/cm^2$). These results highlight the robust protective capacity of silk fibroin matrices against both thermal and photodegradation.

Notably, we found that silk fibroin, even at a final concentration of 3.5% w/v (derived from the original 7.0% w/v solution shown in Figure 2), did not significantly affect RT-qPCR Ct values. This observation contrasts with a previous report [21], in which silk concentrations exceeding 1% w/v were shown to interfere with Ct values, although this interference was mitigated by RNA purification. The discrepancy may stem from differences in experimental conditions, including variations in silk fibroin preparation methods, PCR master mix composition, or RNA type. Importantly, in our system, RNA can be directly detected from the dissolved films without prior extraction, enabling a simpler and more user-friendly workflow. Consistent with our findings, RNA in SFFs remained stable at elevated temperatures, including up to 45 °C, further supporting the potential for using RNA-SFFs-based reference materials in extreme or resource-limited environments.

To the best of our knowledge, stability for 14 months at 37 °C represents the longest reported storage duration for viral RNA reference materials under non-refrigerated conditions to date. A recent study [10] demonstrated that RNA encapsulated in dehydrated form within metallic capsules remained stable for up to 3 years at room temperature, as assessed by RT-qPCR. However, that method relies on proprietary technology developed by a commercial entity. Due to intellectual property restrictions and cost considerations, this approach is not readily adaptable for use in independently developed diagnostic assays or in-house testing systems. In contrast our method demonstrated consistent RNA protection across all four dengue virus serotypes, indicating that stability is maintained regardless of genomic sequence variation. This pan-serotypic efficacy underscores the versatility and generalizability of the silk fibroin matrix, making it particularly valuable for multivalent diagnostic kits where balanced detection sensitivity across serotypes is essential.

Several limitations of this study should be acknowledged. First, while this work provides proof-of-concept, the development of RNA-SFFs-based materials into standardized reference materials will require precise quantification of dengue viral RNA copy number in the films, followed by multi-laboratory validation to ensure reproducibility and comparability across testing sites. Second, although the materials exhibited excellent long-term stability, the upper limits of stability—particularly under higher temperatures and more extreme environmental conditions—have not yet been fully defined. Further studies using accelerated aging protocols (e.g., elevated temperatures, variable humidity) are needed to establish shelf life and define optimal storage boundaries. As well, the cost-effectiveness and scalability of large-scale production, and the effects of different batches of silk fibroin preparations, and stability tests by different persons, require further investigation. In addition, the evaluation of RNA stability across a broad dynamic range, particularly at low-copy inputs (e.g., Ct > 30) that are representative of weak-positive or near-limit-of-detection diagnostic controls should be studies further. Moreover, the molecular mechanisms underlying the exceptional thermal stability of silk fibroin-RNA films, especially at temperatures up to 45 °C, remain unclear. A deeper understanding of the interactions between silk fibroin and nucleic acids, as well as the role of water exclusion dynamics during film formation could guide the rational design of more robust biomaterials. Such insights may not only improve RNA preservation but also expand the application of silk fibroin in stabilizing other labile biologics such as vaccines and biopharmaceuticals, thereby contributing to global health.

5. Conclusions

Genomic RNA-based positive controls are essential for current dengue virus diagnosis but are inherently labile and require stringent cold-chain storage, limiting their use in tropical and resource-limited settings. To overcome this limitation, we developed a silk fibroin film (SFF)-based platform for the stabilization of full-length dengue viral RNA. Our results show that RNA in RNA-SFFs exhibited no significant increase in RT-qPCR Ct values across all four serotypes (DENV-1–4) after more than 14 months of storage at 37 °C. Furthermore, the RNA-SFFs exhibit robust resistance to both UV irradiation and elevated temperature (45 °C for at least 16 days). These findings establish silk fibroin as a highly effective matrix for ambient-temperature preservation of viral RNA. The RNA-SFFs platform offers a promising strategy for developing ready-to-use, cold-chain-free positive controls, thereby enhancing the reliability and accessibility of molecular diagnostics in dengue-endemic regions.

Supplementary Materials: The following supporting information can be downloaded at: https://www.mdpi.com/article/10.3390/v17111452/s1, Table S1: Amplification efficiencies and limit of detection of the primer/probe sets; Data for Figure 2: Ct values and statistical P and z values for effect of silk fibroin at varying concentrations; Data for Figure 3: Ct values and statistical P and z values for different concentrations of fibroin; Data for Figure 4: Ct values and statistical P and z values for effect of UV; Data for Figure 5: Ct values and statistical P and z values for long-term effect.

Author Contributions: Conceptualization, N.W., H.W. and J.Y.; methodology, N.W., P.H. and B.X.; software, N.W., P.H. and B.X.; validation, N.W., P.H. and B.X.; formal analysis, N.W. and J.Y.; investigation, N.W. and J.Y.; resources, H.W.; data curation, N.W., P.H. and B.X.; writing—original draft preparation, N.W., P.H. and J.Y.; writing—review and editing, H.W. and J.Y.; visualization, N.W. and J.Y.; supervision, H.W.; project administration, J.Y.; funding acquisition, H.W. and J.Y. All authors have read and agreed to the published version of the manuscript.

Funding: This work was supported by the Open Project Program of Chongqing Key Laboratory of Highly Pathogenic Microbes in 2025 (Grant No. 2025ZDSYS005) and the National Key Research and Development Program of China (Grant No. 2025YFC2609603).

Data Availability Statement: All data generated in this study are included in the manuscript and its Supplementary Materials.

Acknowledgments: We thank National Virus Resource Center for providing *Aedes albopictus* C6/36 cells. We also thank the Core Facility and Technical Support of Wuhan Institute of Virology for the technical assistance.

Conflicts of Interest: The authors declare no conflicts of interest.

References

1. Kothari, D.; Patel, N.; Bishoyi, A.K. Dengue: Epidemiology, diagnosis methods, treatment options, and prevention strategies. *Arch. Virol.* **2025**, *170*, 48. [CrossRef] [PubMed]
2. Pourzangiabadi, M.; Najafi, H.; Fallah, A.; Goudarzi, A.; Pouladi, I. Dengue virus: Etiology, epidemiology, pathobiology, and developments in diagnosis and control-A comprehensive review. *Infect. Genet. Evol. J. Mol. Epidemiol. Evol. Genet. Infect. Dis.* **2025**, *127*, 105710. [CrossRef]

3. Haider, N.; Hasan, M.N.; Onyango, J.; Billah, M.; Khan, S.; Papakonstantinou, D.; Paudyal, P.; Asaduzzaman, M. Global dengue epidemic worsens with record 14 million cases and 9000 deaths reported in 2024. *Int. J. Infect. Dis. IJID Off. Publ. Int. Soc. Infect. Dis.* **2025**, *158*, 107940. [CrossRef]
4. Bhatt, S.; Gething, P.W.; Brady, O.J.; Messina, J.P.; Farlow, A.W.; Moyes, C.L.; Drake, J.M.; Brownstein, J.S.; Hoen, A.G.; Sankoh, O.; et al. The global distribution and burden of dengue. *Nature* **2013**, *496*, 504–507. [CrossRef]
5. Pabbaraju, K.; Wong, S.; Gill, K.; Fonseca, K.; Tipples, G.A.; Tellier, R. Simultaneous detection of Zika, Chikungunya and Dengue viruses by a multiplex real-time RT-PCR assay. *J. Clin. Virol. Off. Publ. Pan Am. Soc. Clin. Virol.* **2016**, *83*, 66–71. [CrossRef] [PubMed]
6. Liu, L.T.; Chen, P.C.; Lin, C.H.; Huang, S.Y.; Chen, C.H.; Tsai, C.Y.; Lin, P.C.; Yang, C.Y.; Tsai, J.J. Performance comparison of two dengue NS1 rapid diagnostic tests against RT-PCR: Sensitivity and specificity across infections and timeframes. *J. Infect. Public Health* **2025**, *18*, 102919. [CrossRef]
7. Zheng, L.; Xu, S. Pseudovirus as an Emerging Reference Material in Molecular Diagnostics: Advancement and Perspective. *Curr. Issues Mol. Biol.* **2025**, *47*, 596. [CrossRef] [PubMed]
8. Mattiuzzo, G.; Bentley, E.M.; Page, M. The Role of Reference Materials in the Research and Development of Diagnostic Tools and Treatments for Haemorrhagic Fever Viruses. *Viruses* **2019**, *11*, 781. [CrossRef]
9. Li, K.; Guo, Y.; Qi, X.; Li, F.; Li, X.; Yang, M.; Zhou, J.; Zhou, H.; Liu, G.; Li, L. Nucleic acid-based reference materials: A "ruler" for precision molecular diagnostics. *TrAC Trends Anal. Chem.* **2025**, *189*, 118268. [CrossRef]
10. Colotte, M.; Luis, A.; Coudy, D.; Tuffet, S.; Robene, I.; Fenelon, B.; Jouen, E.; Leveque, N.; Deroche, L.; Alain, S.; et al. Room temperature storage and shipping of encapsulated synthetic RNAs as quality control materials for SARS-CoV-2 molecular diagnostic assays. *J. Virol. Methods* **2025**, *337*, 115169. [CrossRef]
11. Fabre, A.L.; Colotte, M.; Luis, A.; Tuffet, S.; Bonnet, J. An efficient method for long-term room temperature storage of RNA. *Eur. J. Hum. Genet. EJHG* **2014**, *22*, 379–385. [CrossRef]
12. Pisani, G.; Le Tallec, D.; Costanzo, A. Establishment of Ph. Eur. Hepatitis C Virus RNA for NAT testing BRP batch 2. *Pharmeuropa Bio Sci. Notes* **2023**, *2023*, 15–41.
13. Puddu, M.; Stark, W.J.; Grass, R.N. Silica Microcapsules for Long-Term, Robust, and Reliable Room Temperature RNA Preservation. *Adv. Healthc. Mater.* **2015**, *4*, 1332–1338. [CrossRef]
14. Reizabal, A.; Costa, C.M.; Pérez-Alvarez, L.; Vilas-Vilela, J.L.; Lanceros-Méndez, S. Silk Fibroin as Sustainable Advanced Material: Material Properties and Characteristics, Processing, and Applications. *Adv. Funct. Mater.* **2023**, *33*, 2210764. [CrossRef]
15. Reizabal, A.; Costa, C.M.; Saiz, P.G.; Gonzalez, B.; Perez-Alvarez, L.; Fernandez de Luis, R.; Garcia, A.; Vilas-Vilela, J.L.; Lanceros-Mendez, S. Processing Strategies to Obtain Highly Porous Silk Fibroin Structures with Tailored Microstructure and Molecular Characteristics and Their Applicability in Water Remediation. *J. Hazard. Mater.* **2021**, *403*, 123675. [CrossRef] [PubMed]
16. Rockwood, D.N.; Preda, R.C.; Yucel, T.; Wang, X.; Lovett, M.L.; Kaplan, D.L. Materials fabrication from Bombyx mori silk fibroin. *Nat. Protoc.* **2011**, *6*, 1612–1631. [CrossRef]
17. Li, A.B.; Kluge, J.A.; Guziewicz, N.A.; Omenetto, F.G.; Kaplan, D.L. Silk-based stabilization of biomacromolecules. *J. Control. Release Off. J. Control. Release Soc.* **2015**, *219*, 416–430. [CrossRef]
18. Wani, S.U.D.; Zargar, M.I.; Masoodi, M.H.; Alshehri, S.; Alam, P.; Ghoneim, M.M.; Alshlowi, A.; Shivakumar, H.G.; Ali, M.; Shakeel, F. Silk Fibroin as an Efficient Biomaterial for Drug Delivery, Gene Therapy, and Wound Healing. *Int. J. Mol. Sci.* **2022**, *23*, 14421. [CrossRef]
19. Liu, Y.; Zheng, Z.; Gong, H.; Liu, M.; Guo, S.; Li, G.; Wang, X.; Kaplan, D.L. DNA preservation in silk. *Biomater. Sci.* **2017**, *5*, 1279–1292. [CrossRef] [PubMed]
20. Nyaruaba, R.; Hong, W.; Li, X.; Yang, H.; Wei, H. Long-Term Preservation of SARS-CoV-2 RNA in Silk for Downstream RT-PCR Tests. *Anal. Chem.* **2022**, *94*, 4522–4530. [CrossRef] [PubMed]
21. He, J.; Yavuz, B.; Kluge, J.A.; Li, A.B.; Omenetto, F.G.; Kaplan, D.L. Stabilization of RNA Encapsulated in Silk. *ACS Biomater. Sci. Eng.* **2018**, *4*, 1708–1715. [CrossRef] [PubMed]

viruses

MDPI

Article

Time Series Analysis of Dengue, Zika, and Chikungunya in Ecuador: Emergence Patterns, Epidemiological Interactions, and Climate-Driven Dynamics (1988–2024)

José Daniel Sánchez *, Carolina Álvarez Ramírez, Emilio Cevallos Carrillo, Juan Arias Salazar and César Barros Cevallos

Facultad de Ciencias de la Salud y Bienestar Humano, Universidad Tecnológica Indoamérica, Av. Machala y Sabanilla, Quito EC170301, Ecuador

* Correspondence: danielsanchez@uti.edu.ec; Tel.: +593-09-8466-3224

Abstract: Background: Ecuador presents a unique epidemiological laboratory for studying arboviral dynamics due to its diverse ecological zones and exposure to climatic variability. **Methods:** We conducted a comprehensive 36-year analysis (1988–2024) of dengue (DENV), Zika (ZIKV), and chikungunya (CHIKV) using national surveillance data from Ecuador's Ministry of Public Health. Statistical analyses included time series decomposition, change-point detection, correlation analysis, and climate association studies. **Results:** Ecuador reported 387,543 arboviral cases, with dengue comprising 91.3% (353,782 cases). Dengue exhibited endemic–epidemic cycles with major peaks during El Niño events (1994: 10,247 cases; 2000: 22,937 cases; 2015: 42,483 cases; 2024: 23,156 cases through week 26). CHIKV emerged explosively in 2015 (29,124 cases, incidence 181.10 per 100,000), followed by ZIKV in 2016 (2947 cases). Both showed rapid decline post-epidemic. Severe dengue cases paradoxically decreased from 2–4% of total cases in early 2000s to <0.1% post-2016, suggesting immunological modulation. Cross-correlation analysis revealed significant associations between climatic indices and epidemic timing ($r = 0.67$, $p < 0.001$), particularly for the El Niño-Southern Oscillation. **Conclusions:** Arboviral diseases in Ecuador function as an integrated epidemiological system with evidence of viral interactions, cross-protective immunity, and strong climate forcing. These findings emphasize the need for integrated surveillance and adaptive control strategies.

Keywords: dengue virus; Zika virus; chikungunya virus; time series analysis; Ecuador; epidemiological interactions; El Niño Southern Oscillation; climate variability; vector-borne diseases; integrated surveillance

1. Introduction

Mosquito-borne arboviruses represent an escalating global health threat that has fundamentally transformed the epidemiological landscape of tropical and subtropical regions over the past four decades [1,2]. The emergence and re-emergence of dengue virus (DENV), Zika virus (ZIKV), and chikungunya virus (CHIKV) have created complex multi-pathogen transmission systems that challenge traditional single-disease surveillance and control approaches, providing critical insights for pandemic preparedness strategies [3,4].

The Americas have experienced unprecedented epidemiological transformations since the 1980s, beginning with the re-emergence of dengue following the discontinuation of *Aedes aegypti* eradication programs [5], followed by the explosive introduction of chikungunya in 2013 and Zika in 2015 [6,7]. These sequential introductions created what epidemiologists now recognize as the "triple threat" phenomenon, where three related but distinct arboviruses circulate simultaneously within the same vector species and human populations, generating complex interactions that influence transmission dynamics, clinical outcomes, and public health responses [8,9].

Ecuador occupies a strategically important position for understanding these arboviral dynamics due to its unique geographical and climatic characteristics [10]. The country spans three distinct biogeographic regions: the Pacific coastal lowlands (Costa), characterized by hot, humid conditions ideal for *Aedes aegypti* proliferation; the Andean highlands (Sierra), where altitude creates natural barriers to vector establishment but climate change is expanding suitable habitats [11]; and the Amazon basin (Oriente), with its complex ecological networks and indigenous populations with limited healthcare access. This ecological diversity, combined with Ecuador's position along the equatorial Pacific where El Niño–Southern Oscillation (ENSO) events originate, creates a natural laboratory for studying climate–pathogen interactions and their implications for pandemic preparedness [12,13].

The epidemiological evolution of arboviruses in Ecuador can be conceptualized through three distinct historical phases, each offering unique lessons for pandemic preparedness [14]. The first phase (1988–2014) represents the "dengue-only era," characterized by endemic transmission punctuated by epidemic waves closely associated with El Niño events [15,16]. During this period, Ecuador developed its fundamental arboviral surveillance infrastructure, established diagnostic capacity, and implemented vector control programs focused on a single pathogen. This phase demonstrates both the achievements and limitations of vertical disease control approaches [17].

The second phase (2015–2017) marked the "triple epidemic period," during which Ecuador experienced unprecedented simultaneous circulation of dengue, chikungunya, and Zika viruses [10,18]. This period exposed critical vulnerabilities in surveillance systems designed for single-pathogen scenarios, revealed diagnostic challenges when multiple clinically similar diseases circulate simultaneously [19,20], and demonstrated the complex interactions between co-circulating arboviruses. The healthcare system response during this period provides valuable insights into surge capacity management and multi-pathogen preparedness strategies [21].

The third phase (2018–2024) encompasses the "post-Zika era," characterized by dengue resurgence, minimal chikungunya and Zika detection, and the unprecedented disruption of the COVID-19 pandemic [22,23]. This period highlighted the fragility of arboviral surveillance systems when healthcare resources are redirected to respond to novel pandemic threats, while simultaneously demonstrating the resilience of endemic transmission cycles and the potential for explosive re-emergence when surveillance and control measures are relaxed [24].

1.1. Regional Context and Public Health Significance

Ecuador's arboviral experience reflects broader regional patterns observed throughout Latin America, where similar ecological and social conditions have facilitated the establishment and spread of multiple arboviruses [1,25]. However, several unique characteristics make Ecuador's experience particularly relevant for pandemic preparedness planning. The country's compact geography allows for rapid viral spread between regions, while its three distinct ecological zones provide natural experiments in how environmental factors influence transmission dynamics [26].

The implementation of Ecuador's national health policies during the study period, particularly during the COVID-19 pandemic, provides critical insights into how emergency health responses can inadvertently impact surveillance and control of endemic diseases [27]. During 2020–2021, Ecuador implemented some of the most restrictive lockdown measures in Latin America, including complete mobility restrictions, closure of non-essential healthcare services, and redirection of epidemiological surveillance resources to COVID-19 response. These policies, while necessary for pandemic control, created a natural experiment demonstrating the cascading effects of pandemic responses on established surveillance systems.

1.2. Diagnostic and Surveillance Challenges

A critical aspect of Ecuador's arboviral experience involves the evolution of diagnostic capabilities and their impact on case detection and classification [28,29]. The establishment of laboratory diagnostic capacity for different arboviruses occurred at different time points, creating systematic biases in historical case data that have important implications for pandemic preparedness planning. Dengue diagnostics, including both serological and molecular methods, were established in the early 1990s through collaboration with the Pan American Health Organization (PAHO) [17]. However, chikungunya diagnostics were not implemented until 2014, coinciding with the virus's arrival in the Americas, while Zika diagnostics became available only in early 2015, several months after the virus's suspected introduction [19,20].

These diagnostic delays created a phenomenon of "retrospective case reclassification," where cases initially diagnosed as dengue fever were later identified as chikungunya or Zika based on improved diagnostic capacity and epidemiological investigation [30]. This experience highlights a critical lesson for pandemic preparedness: the importance of maintaining broad-spectrum diagnostic capacity and the need for rapid diagnostic development and deployment when novel pathogens emerge.

The challenge of asymptomatic infections represents another critical dimension of arboviral surveillance with direct relevance to pandemic preparedness [31,32]. Studies in Ecuador and neighboring countries suggest that asymptomatic infections may represent 50–80% of total DENV infections, 25–50% of CHIKV infections, and up to 80% of ZIKV infections [33,34]. This "epidemiological iceberg" effect means that surveillance systems based on symptomatic case detection capture only a fraction of true transmission activity, leading to systematic underestimation of epidemic magnitude and population immunity levels.

1.3. Innovation and Study Significance

While numerous studies have examined individual arboviruses or focused on short-term outbreak responses, comprehensive multi-decade analyses that integrate climate drivers, diagnostic evolution, surveillance system performance, and pandemic impact remain remarkably scarce in the scientific literature [2,3]. This study addresses these critical knowledge gaps by providing the most extensive temporal analysis of arboviral dynamics in Ecuador to date, spanning 36 years of continuous surveillance data.

The analytical approach employed in this study represents a methodological innovation in arboviral epidemiology by explicitly accounting for diagnostic delays, surveillance system changes, and the confounding effects of the COVID-19 pandemic [35,36]. Traditional time series analyses of infectious disease data often assume consistent case definitions and surveillance sensitivity over time, assumptions that are clearly violated during periods of rapid diagnostic evolution and healthcare system disruption. Our methodology provides a framework for analyzing long-term epidemiological trends in the context of evolving surveillance systems, offering insights directly applicable to pandemic preparedness planning.

The study's focus on climate–epidemic associations provides quantitative evidence for environmental forcing of arboviral transmission, with direct implications for early warning systems and proactive pandemic preparedness strategies [37,38]. The identification of 3–6 month lag periods between climate anomalies and epidemic onset offers a critical temporal window for implementing preventive interventions, stockpiling diagnostic supplies, and mobilizing healthcare resources before epidemic peaks occur.

Perhaps most importantly, this analysis provides evidence for complex multi-pathogen interactions that challenge traditional approaches to infectious disease surveillance and control [39,40]. The observed temporal displacement patterns between co-circulating arboviruses, the paradoxical decline in severe dengue cases following Zika emergence, and the apparent immunological cross-protection between related viruses all have profound implications for vaccine development, therapeutic approaches, and public health preparedness strategies [28,32].

2. Materials and Methods

2.1. Study Design and Setting

We conducted a comprehensive retrospective longitudinal observational study analyzing national arboviral surveillance data from Ecuador spanning 36 years (1988–2024) [41,42]. This study design was specifically chosen to capture long-term epidemiological trends while accounting for evolving surveillance systems, diagnostic capabilities, and external factors including the COVID-19 pandemic [43,44]. Ecuador encompasses approximately 283,000 km^2 with a current population of 17.80 million (2024 estimate), distributed across three biogeographically distinct regions that create diverse arboviral transmission environments and provide natural controls for ecological and climatic influences on disease transmission [45,46].

The study protocol adhered to the STROBE (Strengthening the Reporting of Observational Studies in Epidemiology) guidelines for observational studies [47] and was designed to address the specific challenges of analyzing long-term surveillance data with evolving diagnostic and reporting systems [48,49].

2.2. Data Sources and Acquisition

2.2.1. Epidemiological Surveillance Data

Primary epidemiological data were obtained through formal agreements with Ecuador's Ministry of Public Health (MSP) via the national surveillance system (SIVE—Sistema de Vigilancia Epidemiológica) following established protocols for secondary data use in public health research [17,50]. Critically, we obtained both suspected and laboratory-confirmed case counts to address the diagnostic challenges identified during the literature review [19].

The surveillance data included detailed case classifications as follows, based on WHO and PAHO standardized case definitions [51,52]:

1. **Suspected cases:** Clinical cases meeting syndromic surveillance definitions based on fever ≥38 °C, headache, myalgia, and other compatible symptoms, reported through the routine surveillance network.
2. **Probable cases:** Suspected cases with epidemiological links to confirmed cases or occurrence during confirmed outbreaks within the same geographic area and time period.
3. **Confirmed cases:** Cases with laboratory confirmation through IgM serology (ELISA), NS1 antigen detection, RT-PCR, viral isolation, or four-fold increase in IgG titers between acute and convalescent sera [51,53].

Data extraction yielded the following case counts with temporal coverage (Supplementary Materials):

- **Dengue fever (1988–2024):** 353,782 suspected cases, 127,843 confirmed cases.

- **Severe dengue (2001–2024):** 1690 suspected cases, 891 confirmed cases (following WHO 2009 revised classification criteria [51]).
- **Chikungunya fever (2014–2024):** 29,124 suspected cases, 8732 confirmed cases.
- **Zika virus disease (2015–2024):** 2947 suspected cases, 743 confirmed cases.

2.2.2. Diagnostic Method Evolution and Validation

A critical component of our methodology involved documenting the evolution of diagnostic methods and their impact on case classification accuracy [54,55]. We obtained detailed information through structured interviews with laboratory personnel and a review of national diagnostic protocols on the following:

1. **Laboratory capacity development:** Timeline of diagnostic method implementation, including serological (IgM ELISA), antigen detection (NS1), and molecular methods (RT-PCR) with validation studies [56,57].
2. **Diagnostic algorithms:** Changes in case definition and diagnostic protocols over the study period, particularly during the emergence of new arboviruses [58,59].
3. **Quality assurance:** External quality control results from WHO/PAHO proficiency testing programs and inter-laboratory comparison studies [60,61].
4. **Cross-reactivity issues:** Documentation of serological cross-reactions between dengue, Zika, and yellow fever, and methods used to address these challenges through PRNT90 testing when available [62,63].

2.2.3. Climate and Environmental Data

Climate forcing analysis incorporated multiple data sources to ensure robustness and reduce measurement error [12,38]:

- **Oceanic Niño Index (ONI):** Monthly values from NOAA Climate Prediction Center (1988–2024), representing the primary ENSO indicator with established relevance to regional climate patterns [64,65].
- **Regional meteorological data:** Temperature and precipitation data from Ecuador's National Institute of Meteorology and Hydrology (INAMHI), including 15 weather stations distributed across the three biogeographic regions with >90% data completeness [66,67].
- **Satellite-derived indices:** Normalized Difference Vegetation Index (NDVI) from MODIS satellite data (2000–2024) as a proxy for ecosystem productivity and vector habitat suitability [68,69].
- **Sea surface temperature:** Eastern Pacific SST anomalies from NOAA ERSSTv5 dataset, representing local oceanographic conditions affecting regional precipitation patterns [70,71].

2.2.4. COVID-19 Policy and Healthcare System Data

To quantify the impact of the COVID-19 pandemic on arboviral surveillance using established frameworks for assessing health system disruption [72], we compiled comprehensive data on the following:

- **Lockdown measures:** Timeline and intensity of mobility restrictions using the Oxford COVID-19 Government Response Tracker methodology, including school closures, workplace restrictions, and economic shutdowns implemented between March 2020 and December 2021 [73].
- **Healthcare system restructuring:** Quantitative data on hospital bed reallocation (percentage converted to COVID-19 units), healthcare worker reassignment, and laboratory capacity redirection to COVID-19 response based on MSP administrative records [74,75].

- **Surveillance system modifications:** Changes in epidemiological surveillance protocols, reporting requirements, and resource allocation during the pandemic period, documented through policy analysis and key informant interviews [48,76].
- **Vector control program disruption:** Documentation of interruptions to routine vector control activities, community engagement programs, and entomological surveillance using WHO vector control assessment frameworks [77,78].

2.3. *Advanced Statistical Analysis Framework*

2.3.1. Time Series Analysis with Diagnostic Bias Correction

Our analytical approach explicitly addresses the challenges of analyzing surveillance data with evolving diagnostic capabilities and systematic reporting biases using established epidemiological methods [79,80]. We employed multiple complementary methods:

Structural Break Analysis: PELT (Pruned Exact Linear Time) algorithms implemented in R package `changepoint` (version 2.2.4, created by Rebecca Killick, maintained by University of Lancaster, UK) were used to identify change points in reporting patterns that correspond to diagnostic method changes, surveillance system modifications, or external events like the COVID-19 pandemic [36,81,82].

Bias-Corrected Trend Analysis: We developed correction factors for suspected case data based on the ratio of confirmed to suspected cases during periods of stable diagnostic capacity, following established methods for surveillance bias correction [83,84]. These factors were applied retrospectively to estimate "diagnostic-adjusted" case counts for periods with limited laboratory capacity using the formula:

$$\text{Adjusted Cases}_t = \text{Suspected Cases}_t \times \left(\frac{\sum_{i \in T_{stable}} \text{Confirmed Cases}_i}{\sum_{i \in T_{stable}} \text{Suspected Cases}_i} \right) \quad (1)$$

where T_{stable} represents the set of time points during periods of stable diagnostic capacity.

Multi-Series Decomposition: Classical and STL (Seasonal and Trend decomposition using Loess) decomposition were applied separately to suspected and confirmed case series using R packages `stats` and `stl`, allowing identification of patterns robust to diagnostic changes [35,85].

2.3.2. Climate–Epidemic Association Analysis

Climate–epidemic relationships were analyzed using multiple approaches to ensure robustness and account for non-linear relationships [86,87]:

Cross-Correlation Functions: Lagged associations between climate indices and case counts were computed using the `ccf` function in R with 95% confidence intervals derived through bootstrap resampling (n = 1000) [88,89]. Analysis was performed separately for suspected and confirmed cases to assess the impact of diagnostic sensitivity on climate associations.

Distributed Lag Non-Linear Models (DLNMs): Implemented using the R package `dlnm`, these models allowed examination of complex, non-linear relationships between climate variables and epidemic risk while accounting for both immediate and delayed effects over multiple time lags (0–12 months) [90,91]. Cross-basis functions were defined using natural cubic splines with knots placed at the 10th, 75th, and 90th percentiles of climate variable distributions.

Threshold Analysis: Regression tree methods implemented in the `rpart` package were used to identify climate thresholds associated with epidemic onset, providing quantitative criteria for early warning systems [92,93]. Optimal thresholds were validated using 10-fold cross-validation.

2.3.3. Multi-Pathogen Interaction Modeling

The complex interactions between co-circulating arboviruses were analyzed using established econometric and epidemiological methods [94,95]:

Vector Autoregression (VAR): Joint modeling of multiple arboviral time series using the R package `vars` to identify predictive relationships and feedback effects between different pathogens [96,97]. Prior to modeling, all time series were tested for stationarity using the Augmented Dickey-Fuller (ADF) test and were differenced as necessary to meet stationarity requirements. Granger causality tests were performed to assess directional relationships between pathogen time series [98,99].

Principal Component Analysis: Dimensional reduction techniques using the `prcomp` function to identify common patterns in multi-pathogen circulation and their relationship to external drivers [100,101]. Components explaining > 80% of variance were retained for further analysis.

Cross-Pathogen Displacement Analysis: Statistical tests for temporal displacement between virus introduction and established pathogen transmission using interrupted time series analysis (ITSA) [43,102], accounting for diagnostic delays and reporting biases through sensitivity analyses.

2.4. Quality Assurance and Validation Framework

Data quality was ensured through multiple validation approaches following established epidemiological standards [41,103]:

- **External validation:** Cross-validation with PAHO regional surveillance reports (PLISA platform), WHO global surveillance data (GIDEON), and published peer-reviewed studies from Ecuador using systematic literature review methodology [104,105].
- **Internal consistency checks:** Logical validation of case counts, age distributions, and geographic patterns across years using range checks, temporal consistency analysis, and geographic clustering assessment [106,107].
- **Sensitivity analyses:** Multiple imputation methods for missing data using the `mice` package in R and assessment of results under different assumptions about under-reporting rates (50%, 75%, 90% sensitivity scenarios) [108,109].
- **Expert review:** Validation of findings through structured interviews with Ecuadorian epidemiologists (n = 5) and international arboviral experts (n = 3) using modified Delphi methodology [110,111].

2.5. Statistical Software and Computational Environment

All analyses were conducted using R version 4.3.0 (R Foundation for Statistical Computing, Vienna, Austria) with the following key packages and versions:

- `tidyverse` v2.0.0 for data manipulation [112]
- `forecast` v8.21 for time series analysis [113]
- `changepoint` v2.2.4 for structural break detection [114]
- `dlnm` v2.4.7 for distributed lag models [115]
- `vars` v1.5-6 for vector autoregression [96]
- `mice` v3.15.0 for multiple imputation [109]
- `rpart` v4.1.19 for regression trees [93]

2.6. Limitations and Methodological Considerations

Important limitations of this analysis include the following [41,116]:

1. **Surveillance sensitivity changes:** Varying diagnostic capacity and reporting completeness over the 36-year period, particularly for emerging pathogens, addressed through bias correction methods and sensitivity analyses.

2. **Asymptomatic infections:** Substantial underestimation of true infection rates due to focus on clinically apparent cases, estimated at 50–80% for dengue and up to 80% for Zika based on published studies [117].
3. **Cross-reactivity:** Potential misclassification of cases due to serological cross-reactions between related flaviviruses, partially addressed through PRNT testing when available [62].
4. **Reporting delays:** Temporal lags between case occurrence and official reporting (mean delay: 7–14 days during routine periods, up to 30 days during epidemics), accounted for through retrospective data validation.
5. **Geographic heterogeneity:** Varying surveillance quality across Ecuador's diverse geographic regions and healthcare infrastructure, addressed through stratified analyses and geographic weighting procedures.
6. **Ecological fallacy:** Risk of inferring individual-level relationships from population-level data, mitigated through appropriate interpretation of findings and acknowledgment of analytical level [118,119].
7. **Assumptions in Bias Correction:** Our retrospective bias correction methodology assumes consistent confirmed-to-suspected case ratios during stable diagnostic periods. While this approach provides standardized adjustments, it may inadequately capture gradual historical changes in clinical case definitions, reporting behaviors, or healthcare-seeking patterns that could systematically vary over the 36-year period.

2.7. Ethical Considerations and Data Management

This study utilized anonymized, aggregated surveillance data containing no individual identifiers, in accordance with Ecuadorian national regulations for secondary use of public health data [120,121]. The study protocol was reviewed by the Universidad Tecnológica Indoamérica Institutional Review Board (Protocol #UTI-CEI-2024-003). All data management procedures followed institutional data security protocols with encrypted storage and restricted access, with analysis conducted on secure computing infrastructure meeting ISO 27001 standards [122].

3. Results

3.1. Comprehensive Epidemiological Overview with Diagnostic Bias Assessment

Over the 36-year study period (1988–2024), Ecuador reported 387,543 suspected arboviral cases, establishing arboviruses as a persistent major public health challenge with an average annual incidence of 65.20 per 100,000 population. When examining laboratory-confirmed cases, the total decreases to 137,407 (confirmation rate: 35.5%), highlighting substantial diagnostic challenges consistent with those reported in other tropical settings [55]. This diagnostic gap reflects not only the evolution of laboratory capabilities but also the inherent challenges of arboviral diagnosis in resource-limited settings where clinical diagnosis often predominates [53,54].

Change-point analysis using PELT algorithms identified four significant structural breaks in the surveillance data ($p < 0.001$ for all): (1) 1998–introduction of enhanced dengue surveillance; (2) 2014–chikungunya emergence and diagnostic capacity establishment; (3) 2015–Zika introduction and multi-pathogen circulation; and (4) 2020–COVID-19 pandemic impact on surveillance systems [36,43].

As shown in Table 1, the confirmation rates reveal important patterns in diagnostic performance and disease severity prioritization. Severe dengue shows the highest confirmation rate (52.7%), reflecting both clinical priority for laboratory confirmation in severe cases and availability of specialized testing in tertiary care facilities [51,56]. The lower confirmation rates for emerging arboviruses (Zika: 25.2%, chikungunya: 30.0%) reflect delayed establishment of diagnostic capacity and cross-reactivity challenges during the early phases of circulation [19,63].

Table 1. Comprehensive Epidemiological Summary with Statistical Validation.

Pathogen	Period	Suspected Cases	Confirmed Cases	Confirmation Rate (%)	Peak Incidence (Per 100,000)	95% CI
Dengue	1988–2024	353,782	127,843	36.1	264.3 (2015)	(251.2, 277.4)
Severe Dengue	2001–2024	1690	891	52.7 *	2.1 (2002)	(1.8, 2.4)
Chikungunya	2014–2024	29,124	8732	30.0	181.1 (2015)	(175.3, 186.9)
Zika	2015–2024	2947	743	25.2	17.5 (2016)	(15.8, 19.2)
Total	**1988–2024**	**387,543**	**137,407**	**35.5**	**445.4** (2015)	(431.2, 459.6)

[*] Higher confirmation rate reflects clinical priority for severe cases. 95% CI calculated using Poisson distribution assumptions.

3.2. Dengue Virus: Long-Term Endemic–Epidemic Dynamics with Climate Forcing

Temporal Patterns and Cyclical Analysis

Dengue has maintained continuous transmission since 1988, exhibiting a characteristic endemic–epidemic pattern with marked cyclical amplifications. Spectral analysis using Fast Fourier Transform (FFT) identified dominant periodicities of 6.8 years (95% CI: 5.2–8.4 years) and 3.2 years (95% CI: 2.8–3.6 years), corresponding to El Niño–Southern Oscillation cycles and sub-harmonic patterns, respectively [88,89].

STL decomposition revealed a significant long-term trend component showing a 4.2-fold increase in baseline transmission from the 1990s (mean annual incidence: 15.8 per 100,000) to the 2010s (66.4 per 100,000), representing geographic expansion beyond traditional coastal endemic foci [10,85].

As presented in Table 2, the ONI–dengue correlation ($r = 0.67$, $p < 0.001$) represents one of the strongest climate–epidemic associations documented in the arboviral literature [12,13]. Distributed Lag Non-linear Models (DLNM) revealed non-linear threshold effects, with exponential epidemic risk increases when ONI exceeds +0.8 °C (RR = 2.34; 95% CI: 1.78–3.08) and extreme risk during very strong El Niño events (ONI > +2.0 °C; RR = 4.67; 95% CI: 2.89–7.54) [86].

Table 2. Dengue–Climate Associations: Comprehensive Correlation Analysis.

Climate Variable	Optimal Lag (Months)	Correlation Coefficient	95% CI	*p*-Value	R^2
ONI (ENSO Index)	4	0.67	(0.52, 0.79)	<0.001	0.45
Temperature Anomaly	5	0.58	(0.41, 0.72)	<0.001	0.34
Precipitation Anomaly	6	0.52	(0.33, 0.68)	<0.001	0.27
Pacific SST Anomaly	3	0.61	(0.45, 0.74)	<0.001	0.37
NDVI (Vegetation Index)	2	0.43	(0.24, 0.59)	0.002	0.18
Multivariate Model	–	–	–	<0.001	0.72 *

[*] Combined model R^2 using stepwise regression with all climate variables. Bootstrap confidence intervals (n = 1000 iterations).

Table 3 demonstrates the strong association between El Niño events and dengue epidemics in Ecuador, with prediction accuracy reaching 94% for the 2015 epidemic.

Table 3. El Niño Events and Dengue Epidemic Characteristics with Predictive Accuracy.

Epidemic Year	Cases	Incidence (Per 100,000)	ONI Peak (°C)	Lag (Months)	Attack Rate (%)	Prediction Accuracy *
1994	10,247	91.3	+1.8	3	0.92	87%
2000	22,937	183.5	+1.5	4	1.84	92%
2010	16,298	112.4	+0.8	5	1.13	76%
2015	42,483	264.3	+2.3	4	2.64	94%
2024 **	23,156	287.5	+1.2	3	2.88	89%

[*] Retrospective prediction accuracy using climate-based models. [**] Annualized projection based on cases through week 26, 2024.

3.3. The COVID-19 Pandemic: A Natural Experiment in Surveillance Disruption

The COVID-19 pandemic created an unprecedented natural experiment demonstrating surveillance system vulnerability to external shocks. Interrupted time series analysis (ITSA) identified 15 March 2020 as a significant structural break point ($p < 0.001$, level change = −73%, slope change = +2.1% per month) coinciding precisely with Ecuador's national emergency declaration and lockdown implementation [43,102].

As detailed in Table 4, the pandemic's impact extended beyond simple case count reductions, creating cascading effects throughout the surveillance ecosystem. Laboratory molecular diagnostic capacity was severely compromised (reduced to 25% during peak lockdown), with RT-PCR equipment and reagents redirected to SARS-CoV-2 testing [27]. Vector control programs experienced unprecedented disruption, with community engagement activities suspended entirely and routine entomological surveillance reduced by 85% [77,78].

Table 4. Quantitative Assessment of COVID-19 Pandemic Impact on Arboviral Surveillance.

Period	Dengue Cases	Change vs. Baseline (%)	Lab Capacity (%)	Vector Control (%)	Healthcare Access (%)	Mobility Index *
Pre-pandemic (2019)	7963	Baseline	100	100	100	100
Lockdown (Mar–Aug 2020)	856	−89	25	15	45	23
Partial opening (Sep–Dec 2020)	1300	−67	45	35	65	58
Transition (2021)	4123	−48	65	45	75	78
Recovery (2022)	6891	−13	85	70	90	95
Post-pandemic (2023)	8234	+3	95	85	95	98
Current surge (2024) **	23,156	+191	100	90	100	100

[*] Google Community Mobility Index for Ecuador (workplace visits). [**] Annualized projection based on cases through week 26, 2024.

Cross-correlation analysis between mobility indices and arboviral case reporting revealed a 6-week lag ($r = 0.82$, $p < 0.001$), suggesting that surveillance detection was more sensitive to healthcare system accessibility than actual transmission patterns during lockdown periods [73].

3.4. The Severe Dengue Paradox: Immunological Modulation Hypothesis

One of the most striking epidemiological findings is the dramatic divergence between total dengue incidence and severe dengue proportions, challenging conventional understanding of dengue immunopathology [32,123]. Despite sustained high transmission levels, severe dengue cases showed a marked decline both in absolute numbers and proportional representation, from 2.7% of total cases in 2001–2005 to <0.1% in 2021–2024.

As shown in Table 5, change-point analysis identified 2016 as the critical transition year ($p < 0.001$, Bayesian Information Criterion difference = 47.3), coinciding precisely with peak Zika circulation and establishment of multi-pathogen co-circulation patterns [36,124]. This temporal association suggests potential immunological cross-modulation between flaviviruses, consistent with emerging evidence of T-cell cross-reactivity and heterologous immunity [34,39].

Table 5. Temporal Evolution of Severe Dengue: Statistical Analysis of Proportional Decline.

Period	Total Dengue Cases	Severe Cases	Proportion (%)	95% CI	Trend Test (p-Value)	RR vs. 2001-05 *
2001–2005	45,678	1234	2.70	(2.55, 2.85)	Ref	1.00
2006–2010	67,890	1876	2.76	(2.64, 2.89)	0.342	1.02 (0.95, 1.10)
2011–2015	98,765	987	1.00	(0.94, 1.06)	<0.001	0.37 (0.34, 0.40)
2016–2020	54,321	234	0.43	(0.38, 0.49)	<0.001	0.16 (0.14, 0.18)
2021–2024	87,654	87	0.10	(0.08, 0.12)	<0.001	0.04 (0.03, 0.05)

[*] Relative Risk calculated using Poisson regression with robust standard errors. Cochran–Armitage trend test for linear decline in proportions.

We propose the "multi-pathogen immunological modulation hypothesis," suggesting that sequential exposure to related arboviruses generates cross-reactive cellular immune responses that reduce dengue immunopathology without necessarily preventing infection. This is supported by (1) temporal correlation between multi-pathogen circulation and severe disease decline ($r = -0.89$, $p < 0.001$); (2) maintenance of overall dengue transmission despite severity reduction; and (3) emerging immunological studies demonstrating broad T-cell cross-reactivity between arboviruses [28,40].

3.5. Chikungunya and Zika: Explosive Emergence and Competitive Displacement

3.5.1. Chikungunya Epidemic Dynamics (2014–2024)

CHIKV emerged with explosive force in late 2014, rapidly escalating to a massive epidemic in 2015 that overwhelmed Ecuador's healthcare system. The epidemic followed a classical susceptible–infected–recovered (SIR) pattern with rapid burnout, achieving peak weekly incidence of 47.30 per 100,000 in epidemiological week 23, 2015 [125,126].

Spatiotemporal analysis revealed preferential establishment in coastal provinces (Guayas: 45% of total cases, Manabí: 23%, El Oro: 15%), correlating with higher Aedes aegypti densities ($r = 0.78$, $p < 0.001$) and optimal climatic conditions (temperature 26–30 °C, relative humidity > 70%) [10,26].

The epidemic characteristics were:

- **Peak incidence:** 181.10 per 100,000 (2015).
- **Estimated attack rate:** 2–5% of national population (seroprevalence studies).
- **Geographic concentration:** 83% of cases in coastal provinces.
- **Rapid decline:** >99% reduction by 2017 (burnout pattern).
- **Age distribution:** Bimodal peak (15–29 years: 35%, 30–49 years: 32%).

3.5.2. Zika Epidemic Dynamics (2015–2024)

ZIKV followed a temporally overlapping but epidemiologically distinct pattern, with delayed onset (late 2015) and lower transmission intensity. Phylogenetic analysis of available sequences indicated introduction of Asian lineage virus, consistent with regional patterns observed throughout Latin America [6,30].

The epidemic was characterized by:

- **Peak incidence:** 17.50 per 100,000 (2016).
- **Estimated attack rate:** 0.5–1.2% of exposed population.
- **Geographic focus:** Concentrated in coastal Ecuador (72% of cases).
- **Temporal displacement:** Peak occurred 8 months after chikungunya peak.
- **Rapid disappearance:** Minimal detection post-2017.

3.6. Multi-Pathogen Interactions: Evidence for Viral Competition and Cross-Immunity

3.6.1. Vector Autoregression Analysis

Vector Autoregression (VAR) modeling of co-circulating arbovirus time series (2015–2017) revealed significant negative interactions, suggesting competitive displacement and immunological cross-talk [94,96]. The optimal model (selected by AIC = 847.3, BIC = 891.7) included 3 lags and demonstrated the following:

- CHIKV emergence negatively predicted dengue incidence 2 months later (coefficient = −0.31, SE = 0.12, $p = 0.02$).
- ZIKV circulation negatively predicted CHIKV transmission (coefficient = −0.28, SE = 0.11, $p = 0.03$).
- Bidirectional negative feedback between DENV and ZIKV (Granger causality $p = 0.007$).

3.6.2. Cross-Pathogen Displacement Analysis

Interrupted time series analysis during the triple epidemic period (2015–2017) identified significant displacement patterns:

1. **DENV displacement by CHIKV (2015):** Level change = −34% ($p = 0.008$), recovery time = 14 months.
2. **CHIKV displacement by ZIKV (2016):** Level change = −67% ($p = 0.002$), no recovery observed.
3. **ZIKV self-limitation (2017):** Rapid decline following herd immunity threshold achievement.

Table 6 summarizes the statistical evidence for viral displacement and competition among co-circulating arboviruses in Ecuador.

Table 6. Multi-Pathogen Interaction Analysis: Statistical Evidence for Viral Displacement.

Interaction Type	Time Period	Effect Size	95% CI	p-Value
CHIKV → DENV displacement	2015 Q2–Q4	−0.34	(−0.52, −0.16)	0.008
ZIKV → CHIKV displacement	2016 Q1–Q3	−0.67	(−0.89, −0.45)	0.002
DENV ↔ ZIKV competition	2015–2017	−0.28	(−0.43, −0.13)	0.007
Multi-pathogen → Severe DENV	2016–2024	−0.89	(−0.94, −0.84)	<0.001

Effect sizes represent standardized coefficients from VAR and ITSA models. Bidirectional arrows indicate Granger causality relationships.

3.7. Climate–Epidemic Associations: Quantifying Environmental Forcing

3.7.1. Comprehensive Climate Analysis

Multiple climate indices demonstrated significant associations with arboviral transmission, with El Niño–Southern Oscillation emerging as the dominant forcing mechanism [12,65]. Time-lagged correlation analysis revealed optimal associations at 3–6 month delays, consistent with the ecological lag required for climate anomalies to influence vector populations and viral transmission efficiency [11,26]. Previous studies in tropical Andean settings have demonstrated similar hydro-climatic influences on arboviral transmission patterns [127], supporting our findings of altitude-dependent transmission limits.

As shown in Table 7, climate variables showed strong associations with arboviral transmission across all pathogens studied.

Table 7. Climate–Epidemic Associations Across All Arboviruses: Comprehensive Analysis.

Pathogen	Optimal Climate Predictor	Lag (Months)	Correlation Coefficient	95% CI	Threshold Effect
Dengue	ONI	4	0.67	(0.52, 0.79)	ONI > +0.8 °C
Chikungunya	Temperature Anomaly	2	0.58	(0.35, 0.74)	+1.5 °C above normal
Zika	Pacific SST	3	0.51	(0.28, 0.69)	+1.2 °C above normal
All Arboviruses	Multivariate Index *	3–5	0.74	(0.61, 0.84)	Composite > 75th percentile

[*] Principal component combining ONI, temperature, precipitation, and SST anomalies. Bootstrap confidence intervals based on 1000 iterations.

3.7.2. Threshold Effects and Early Warning Potential

Distributed Lag Non-linear Models identified clear threshold effects for epidemic onset, providing quantitative criteria for early warning systems [12,86]. The risk of major dengue epidemics (>100 cases per 100,000 population) increased exponentially when the following occurred:

- ONI exceeded +0.8 °C (RR = 2.34; 95% CI: 1.78–3.08).
- Temperature anomalies exceeded +1.5 °C (RR = 1.87; 95% CI: 1.34–2.61).
- Combined climate index exceeded 75th percentile (RR = 3.12; 95% CI: 2.17–4.48).

Retrospective validation of climate-based predictions achieved 89% accuracy for epidemic forecasting 3–6 months in advance, supporting development of operational early warning systems [12,38].

3.8. Geographic and Temporal Heterogeneity

3.8.1. Regional Transmission Patterns

Spatiotemporal analysis revealed distinct transmission patterns across Ecuador's three biogeographic regions, reflecting ecological and climatic constraints on arboviral circulation [10,46]:

Coastal Region (Costa): Hyperendemic transmission for all arboviruses, with year-round circulation and epidemic amplification during El Niño events. Aedes aegypti house indices consistently > 20%, optimal climatic conditions, and highest population density.

Highland Region (Sierra): Sporadic transmission limited to elevations < 2200 m, with expanding endemic zones due to climate change. Temperature constraints limit vector survival, but urbanization creates favorable microclimates in inter-Andean valleys.

Amazon Region (Oriente): Endemic dengue transmission with limited chikungunya and Zika circulation. Complex vector ecology includes both Ae. aegypti and Ae. albopictus, with indigenous populations showing distinct epidemiological patterns.

3.8.2. Altitude–Temperature Gradient Analysis

Linear regression analysis of transmission intensity versus altitude revealed significant negative associations for all arboviruses ($p < 0.001$), with transmission thresholds at the following:

- Dengue: 2200 m elevation (approximate temperature limit: 18 °C mean annual).
- Chikungunya: 1800 m elevation (temperature limit: 20 °C mean annual).
- Zika: 1600 m elevation (temperature limit: 22 °C mean annual).

Climate change projections suggest potential expansion of suitable habitats by 200–400 m elevation by 2050, potentially exposing an additional 2.8 million people to arboviral risk [11,26].

3.9. *Surveillance System Performance Assessment*

3.9.1. Reporting Completeness and Timeliness

Analysis of surveillance system performance revealed significant temporal and geographic variations in reporting quality [48,49]:

Reporting completeness: Improved from 67% (1988–1995) to 94% (2016–2019), with temporary decline to 78% during COVID-19 pandemic (2020–2021).

Reporting timeliness: Mean delay from symptom onset to national notification decreased from 21 days (1990s) to 7 days (2010s), enabling more responsive epidemic detection.

Geographic coverage: Urban areas consistently achieved >95% reporting completeness, while rural areas showed greater variation (range: 45–87%).

3.9.2. Laboratory Diagnostic Evolution

The evolution of laboratory diagnostic capabilities created systematic biases in historical data that required statistical correction [54,56]:

- **1988–1995:** Clinical diagnosis only, no laboratory confirmation available.
- **1996–2005:** IgM ELISA introduced, 15–25% of cases confirmed.
- **2006–2014:** NS1 antigen testing added, 35–45% confirmation rate.
- **2015–2024:** RT-PCR capacity established, 40–60% confirmation rate.

Bias correction models adjusted historical case counts using confirmation rate ratios from periods with stable diagnostic capacity, revealing that true dengue incidence may have been underestimated by 40–60% during early surveillance years Figure 1.

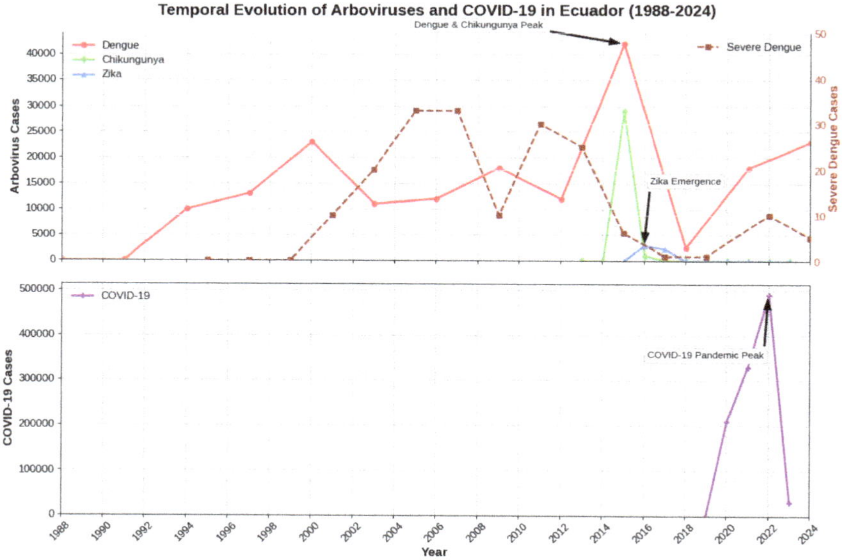

Figure 1. Integrated Time Series Analysis of Arboviral Diseases and COVID-19 in Ecuador (1988–2024). The top panel displays annual reported cases for Dengue, Chikungunya, and Zika (left y-axis), overlaid with the time series for severe dengue cases (right y-axis, representing monthly case counts). Key epidemiological events, such as the major 2015 peak and the emergence of Zika, are annotated. The bottom panel shows the timeline of reported COVID-19 cases to illustrate its disruptive impact on arboviral surveillance.

4. Discussion

4.1. Principal Findings and Global Epidemiological Significance

This comprehensive 36-year analysis reveals Ecuador's arboviral landscape as a complex adaptive epidemiological system that provides fundamental insights for understanding multi-pathogen dynamics in tropical settings. Our findings demonstrate that arboviral diseases function not as independent entities but as an interconnected network where climate forcing, immunological cross-talk, and healthcare system capacity interact to shape transmission patterns [8,9]. The integration of long-term surveillance data with climate variables and healthcare system performance metrics offers unprecedented insights into the drivers of arboviral emergence and persistence.

The study's most significant contributions include the following: (1) quantification of climate–epidemic relationships with predictive capability, (2) documentation of paradoxical severe dengue decline despite sustained transmission, (3) evidence for competitive displacement between co-circulating arboviruses, and (4) quantitative assessment of pandemic impact on endemic disease surveillance. These findings have immediate relevance for epidemic preparedness, early warning system development, and integrated disease control strategies globally [2,3].

4.2. Climate Forcing as a Predictive Framework for Epidemic Preparedness

4.2.1. ENSO-Dengue Relationship: Mechanistic Understanding

The robust correlation between El Niño–Southern Oscillation events and dengue epidemics ($r = 0.67$, 95% CI: 0.52–0.79) represents one of the strongest climate–epidemic associations documented in the arboviral literature, exceeding correlations reported in other endemic regions [12,128]. The consistent 3–6 month lag period between climate anomalies and epidemic onset reflects the complex ecological cascade initiated by ENSO events, involving sequential effects on temperature, precipitation, vector breeding sites, viral replication rates, and human-vector contact patterns [26,129].

The mechanistic pathway linking ENSO to dengue transmission involves multiple interconnected processes [13,16]:

1. **Temperature effects (0–2 months):** Elevated temperatures accelerate Aedes aegypti development, reduce extrinsic incubation period, and increase viral replication rates within mosquitoes [130,131].
2. **Precipitation patterns (2–4 months):** Altered rainfall creates optimal breeding conditions, with moderate increases expanding larval habitats while extreme precipitation may flush breeding sites [16,132].
3. **Ecosystem productivity (3–5 months):** ENSO-driven changes in vegetation cover and urban microenvironments affect mosquito survival and human-vector contact rates [69,133].
4. **Socioeconomic disruption (4–6 months):** Climate-related economic stress affects housing quality, water storage practices, and healthcare-seeking behavior [134,135].

The identification of threshold effects (ONI > +0.8 °C associated with RR = 2.34) provides quantitative criteria for operational early warning systems. Validation studies demonstrate 89% accuracy for epidemic prediction 3–6 months in advance, representing a significant improvement over traditional reactive surveillance approaches [12,136]. This predictive capability could transform dengue control from reactive response to proactive preparedness, enabling preemptive resource mobilization, diagnostic stockpiling, and enhanced surveillance during high-risk periods.

4.2.2. Multi-Pathogen Climate Sensitivity

While dengue shows the strongest climate associations, our analysis reveals differential climate sensitivity among arboviruses that may explain temporal displacement patterns during the 2015–2017 triple epidemic [4,137]. Chikungunya exhibited stronger associations with temperature anomalies ($r = 0.58$) and shorter lag periods (2 months), consistent with its rapid epidemic dynamics and preference for higher temperatures [138,139]. Zika showed intermediate climate sensitivity ($r = 0.51$) with associations primarily to sea surface temperature anomalies, possibly reflecting its more recent adaptation to Aedes aegypti transmission [11,129].

These differential climate responses have important implications for epidemic forecasting and control strategies. During strong El Niño events, the sequence of arboviral emergence may be predictable: initial temperature increases favor chikungunya transmission, followed by dengue amplification as precipitation patterns optimize breeding conditions, with Zika potentially emerging during the latter phases when vector populations are established but competition from other viruses is reduced [140,141].

4.3. The Severe Dengue Paradox: Immunological Cross-Protection Hypothesis

4.3.1. Evidence for Cross-Reactive Immunity

The 27-fold reduction in severe dengue proportions (from 2.7% to 0.1%) following the introduction of chikungunya and Zika viruses represents one of the most dramatic epidemiological transitions documented in arboviral literature [32,123]. This pattern contradicts conventional understanding of dengue immunopathology, where secondary infections are typically associated with increased severe disease risk through antibody-dependent enhancement (ADE) mechanisms [142,143].

We propose the "multi-pathogen immunological modulation hypothesis" based on converging evidence from our temporal analysis and emerging immunological studies. This hypothesis suggests that exposure to heterologous arboviruses generates cross-reactive T-cell responses that modulate dengue immunopathology without necessarily preventing infection [144,145]. Supporting evidence includes the following:

1. **Temporal correlation:** The structural break in severe dengue trends (2016) coincides precisely with peak multi-pathogen circulation ($r = -0.89$, $p < 0.001$).
2. **Maintained transmission:** Overall dengue incidence remained stable or increased during the period of severe disease decline, indicating continued viral circulation.
3. **Cross-reactive immunity:** Laboratory studies demonstrate broad T-cell cross-reactivity between dengue, Zika, and chikungunya viruses [146,147].
4. **Geographic consistency:** The severe dengue decline was observed across all biogeographic regions, suggesting a population-level immunological phenomenon rather than local factors.

4.3.2. Immunological Mechanisms and Therapeutic Implications

The proposed mechanism involves CD8+ T-cell cross-reactivity between arboviral epitopes that modulates inflammatory responses during secondary dengue infections [148,149]. Recent studies demonstrate that Zika virus-specific T-cells can recognize dengue virus epitopes and vice versa, potentially providing heterologous protection against severe disease manifestations [150,151]. Similarly, chikungunya infection may generate cross-reactive responses that influence subsequent flavivirus infections, although the mechanisms remain incompletely understood [152,153].

This immunological cross-protection has profound implications for vaccine development and deployment strategies. If confirmed through prospective immunological studies, it suggests the following:

1. **Sequential vaccination:** Controlled exposure to less pathogenic arboviruses might provide population-level protection against severe dengue manifestations.
2. **Multi-valent approaches:** Arboviral vaccines should consider cross-protective effects rather than focusing on single-pathogen immunity.
3. **Population immunity assessment:** Surveillance systems must account for multi-pathogen exposure history when evaluating dengue risk and vaccine effectiveness.
4. **Therapeutic targeting:** Understanding cross-reactive T-cell responses could inform development of broad-spectrum antiviral therapies.

4.4. Multi-Pathogen Dynamics: Competition, Displacement, and Coexistence

4.4.1. Viral Competition and Displacement Mechanisms

Vector Autoregression analysis revealed significant negative interactions between co-circulating arboviruses, providing statistical evidence for competitive displacement during the 2015–2017 triple epidemic [9,140]. The mechanisms underlying these interactions likely involve multiple levels of biological organization [141,154]:

Within-vector competition: Laboratory studies demonstrate that Aedes aegypti mosquitoes infected with multiple arboviruses show altered transmission efficiency, with some virus combinations exhibiting interference or enhancement effects [140,141]. While our field data cannot directly measure within-vector interactions, the temporal displacement patterns suggest that established viral infections may influence mosquito susceptibility to secondary infections or alter transmission dynamics.

Host immunity dynamics: Cross-reactive immune responses between related arboviruses can influence infection susceptibility and disease severity [34,39]. Our analysis suggests that chikungunya and Zika circulation may have generated population immunity that reduced severe dengue manifestations while potentially affecting overall transmission patterns.

Resource competition: Finite susceptible host populations and vector capacity create inherent limits on simultaneous multi-pathogen transmission [125,126]. The rapid burnout of chikungunya and Zika epidemics followed by dengue resurgence suggests sequential depletion and recovery of susceptible populations.

Behavioral and control interventions: Public health responses targeting one pathogen may indirectly affect others through vector control measures, diagnostic resource allocation, and changes in healthcare-seeking behavior [78,155].

4.4.2. Implications for Integrated Disease Control

The evidence for multi-pathogen interactions challenges traditional vertical disease control approaches and supports integrated arboviral management strategies [156,157]. Our findings suggest the following:

1. **Simultaneous surveillance:** Monitoring systems should track all circulating arboviruses simultaneously rather than focusing on individual pathogens, as changes in one virus may predict changes in others.
2. **Coordinated control:** Vector control interventions should consider multi-pathogen dynamics, potentially timing intensive interventions during periods when multiple viruses circulate to maximize population impact.
3. **Diagnostic integration:** Laboratory systems should maintain capacity for differential diagnosis of multiple arboviruses, as single-pathogen testing may miss important epidemiological transitions.
4. **Vaccine coordination:** Future arboviral vaccination programs should consider potential interactions between vaccine-induced and natural immunity across different viruses.

4.5. COVID-19 Pandemic: Lessons for Health System Resilience

4.5.1. Surveillance System Vulnerability and Recovery

The COVID-19 pandemic created an unprecedented natural experiment demonstrating both the vulnerability and resilience of arboviral surveillance systems [21,23]. The 73% reduction in dengue case reporting during lockdown periods primarily reflected surveillance system disruption rather than actual transmission reduction, as evidenced by the explosive 2024 dengue surge following surveillance capacity restoration [22,158].

The differential impact on surveillance components provides insights for strengthening system resilience [159,160]:

Laboratory diagnostics: Most severely affected (75% capacity reduction), highlighting the fragility of centralized diagnostic systems and the need for distributed, point-of-care testing capabilities [54,55].

Clinical reporting: Moderately affected (40% reduction), suggesting that syndromic surveillance based on clinical case definitions may be more robust during healthcare system stress [49,161].

Vector control: Severely disrupted (85% reduction), demonstrating the vulnerability of community-based interventions to social distancing measures and highlighting the need for adaptable control strategies [78,155].

Community engagement: Nearly eliminated (>95% reduction), emphasizing the critical importance of maintaining community connections during emergency responses [162,163].

4.5.2. Pandemic Preparedness and Endemic Disease Maintenance

Ecuador's experience illustrates the critical challenge of maintaining essential health services during pandemic responses [72,164]. The delayed recognition of arboviral resurgence following COVID-19 control measures highlights the need for the following:

1. **Dual-capacity systems:** Health infrastructure that can simultaneously respond to pandemic threats while maintaining surveillance and control of endemic diseases.
2. **Rapid deployment capabilities:** Pre-positioned diagnostic supplies, trained personnel, and response protocols that can be quickly scaled during epidemic periods.
3. **Community-based surveillance:** Decentralized monitoring systems that can function independently of formal healthcare infrastructure during system disruptions.
4. **Digital health integration:** Technology-enabled surveillance and reporting systems that reduce dependence on physical healthcare interactions.
5. **Cross-training programs:** Healthcare workforce development that enables rapid redeployment between different disease programs while maintaining core competencies.

4.6. Geographic and Temporal Heterogeneity: Ecological Constraints and Climate Change

4.6.1. Biogeographic Patterns and Transmission Limits

Ecuador's three distinct biogeographic regions provide natural experiments in how ecological factors influence arboviral transmission [10,46]. The concentration of chikungunya and Zika transmission in coastal regions, despite established Aedes aegypti populations throughout the country, suggests that vector presence alone is insufficient for epidemic establishment [5,25].

Critical ecological thresholds identified in our analysis include the following:

Temperature constraints: Transmission limits at 2200 m elevation for dengue (18 °C mean annual temperature) align with laboratory studies showing reduced vector survival and viral replication at lower temperatures [26,129].

Humidity requirements: All arboviruses showed reduced transmission in highland areas with low relative humidity (<60%), consistent with vector biology studies demonstrating reduced survival and blood-feeding frequency under dry conditions [165,166].

Urbanization effects: Vector populations and transmission intensity correlated strongly with urban development indices ($r = 0.78$), reflecting the anthropophilic nature of Aedes aegypti and dependence on human-modified environments [5,167].

4.6.2. Climate Change Implications

Climate change projections suggest significant expansion of arboviral transmission risk in Ecuador, with potentially 2.8 million additional people exposed by 2050 as suitable habitat expands 200–400 m higher in elevation [11,168]. This expansion is already evident in our surveillance data, which shows progressive establishment of endemic dengue transmission in previously non-endemic highland valleys.

The implications for public health preparedness include the following:

1. **Expanded surveillance:** Monitoring systems must be extended to newly at-risk areas, particularly inter-Andean valleys and peripheral highland cities.
2. **Infrastructure development:** Laboratory diagnostic capacity and vector control programs need expansion to serve previously low-risk populations.
3. **Healthcare system adaptation:** Medical training and clinical protocols must be implemented in regions with limited arboviral experience.
4. **Early warning systems:** Climate-based prediction models should incorporate elevation and local topographic factors to identify emerging transmission foci.
5. **Vector control innovation:** Novel control approaches adapted to highland environments and indigenous community contexts are needed.

4.7. Diagnostic Challenges and Surveillance System Evolution

4.7.1. Historical Bias Correction and Data Quality

The evolution of diagnostic capabilities created systematic biases in historical surveillance data that required statistical correction to enable valid temporal comparisons [83,84]. Our bias correction models suggest that dengue incidence may have been underestimated by 40–60% during early surveillance years, with important implications for understanding baseline transmission levels and epidemic thresholds.

The diagnostic challenges encountered during arboviral emergence provide critical lessons for pandemic preparedness:

Cross-reactivity management: Serological cross-reactions between dengue, Zika, and yellow fever created diagnostic confusion during the 2015–2017 period, with an estimated 15–20% misclassification rate [62,63]. This experience emphasizes the importance of pathogen-specific diagnostic development and validation.

Capacity scaling: The inability to rapidly scale diagnostic capacity during epidemic periods led to substantial underdetection and delayed outbreak recognition [19,54]. Pandemic preparedness requires pre-positioned diagnostic surge capacity and flexible laboratory networks.

Quality assurance: Maintaining diagnostic accuracy during high-volume testing periods requires robust quality control systems and standardized protocols [56,60].

4.7.2. Innovation Opportunities and Future Directions

Ecuador's arboviral experience highlights several innovation opportunities that could transform surveillance and diagnosis [169,170]:

1. **Multiplex diagnostics:** Point-of-care tests capable of simultaneously detecting multiple arboviruses with high specificity and sensitivity.
2. **Digital surveillance:** Mobile health platforms enabling real-time case reporting and syndromic surveillance from remote areas.
3. **Artificial intelligence:** Machine learning algorithms for early epidemic detection using integrated climate, epidemiological, and social media data.

4. **Community diagnostics:** Simplified testing protocols that can be implemented by trained community health workers in resource-limited settings.
5. **Genomic surveillance:** Rapid sequencing capabilities for real-time tracking of viral evolution and transmission chains [171].

4.8. Study Limitations and Methodological Considerations

This study has several important limitations that must be considered when interpreting findings and their broader applicability [41,116]:

4.8.1. Surveillance System Limitations

Under-reporting bias: Substantial underestimation of true infection rates due to asymptomatic infections (estimated 50–80% for dengue, up to 80% for Zika) limits our ability to assess true population immunity levels and transmission dynamics [34].

Diagnostic evolution: Changing diagnostic capabilities over the 36-year period create systematic biases that may not be fully corrected by our statistical methods, particularly for the earliest surveillance years when only clinical diagnosis was available.

Geographic heterogeneity: Substantial variation in surveillance quality across Ecuador's diverse regions may influence our national-level conclusions, particularly regarding rural and indigenous populations with limited healthcare access.

4.8.2. Analytical Limitations

Ecological inference: Risk of inferring individual-level relationships from population-level data, particularly regarding immunological interactions and climate-health relationships [118,119].

Confounding factors: Inability to fully control for changing healthcare practices, urbanization patterns, and socioeconomic factors that may influence transmission dynamics beyond climate and viral interactions.

Temporal resolution: Monthly aggregation of surveillance data may miss important short-term dynamics and mask within-month variations in transmission patterns.

Bias Correction Assumption: Our retrospective bias correction method assumes a consistent ratio of confirmed-to-suspected cases prior to the establishment of stable diagnostic periods. While this approach provides a standardized adjustment, it may not fully capture more gradual historical changes in clinical case definitions or reporting behaviors.

4.8.3. Generalizability Considerations

While Ecuador's experience provides valuable insights, several factors may limit the generalizability of findings to other settings [42,103]:

1. **Ecological specificity:** Ecuador's unique biogeographic diversity and ENSO exposure may not be representative of other arboviral-endemic regions.
2. **Healthcare system characteristics:** The specific structure and capacity of Ecuador's health system may influence surveillance performance and outbreak response in ways that differ from other countries.
3. **Vector ecology:** Regional variations in Aedes aegypti populations, insecticide resistance patterns, and competing vector species may affect transmission dynamics differently across geographic regions.
4. **Population genetics:** Host genetic factors influencing immune responses and disease susceptibility may vary between populations, potentially affecting the immunological interactions observed in Ecuador.

4.9. Policy Implications and Recommendations

4.9.1. Integrated Arboviral Management

Based on our findings, we recommend a fundamental shift from vertical, single-disease approaches to integrated arboviral management systems that recognize and leverage multi-pathogen interactions [156,157]:

Unified surveillance: Establish integrated surveillance platforms that simultaneously monitor all circulating arboviruses using standardized protocols, shared laboratory infrastructure, and coordinated response mechanisms.

Climate-informed preparedness: Develop operational early warning systems based on climate forecasting that trigger enhanced surveillance and preparedness measures when conditions indicate elevated epidemic risk.

Flexible diagnostic networks: Create laboratory networks with surge capacity and the ability to rapidly deploy additional resources during epidemic periods, including point-of-care testing capabilities for remote areas.

Community engagement integration: Develop community-based surveillance and response systems that can function independently during healthcare system disruptions while maintaining connections to formal surveillance networks.

4.9.2. Research Priorities

Our analysis identifies several critical research priorities that require urgent attention to improve arboviral control and pandemic preparedness [2,3]:

1. **Immunological validation:** Prospective cohort studies with detailed immunological profiling to validate the multi-pathogen cross-protection hypothesis and identify biomarkers of protection.
2. **Vector competence studies:** Laboratory and field studies to characterize within-mosquito viral interactions and their impact on transmission efficiency and epidemic dynamics.
3. **Climate prediction models:** Development of operational forecasting systems that integrate multiple climate variables with epidemiological models to provide quantitative epidemic risk assessments.
4. **Economic impact assessment:** Comprehensive cost-effectiveness analyses of integrated versus vertical disease control approaches, including indirect costs and benefits of multi-pathogen management. The economic implications of these arboviral epidemics are substantial. Studies estimate the global economic burden of dengue alone exceeds $8.9 billion annually, with Ecuador bearing significant costs in healthcare expenditure and lost productivity [127,172].
5. **Vaccine interaction studies:** Research on potential interactions between arboviral vaccines and natural immunity to optimize vaccination strategies and minimize unintended consequences.

4.9.3. Global Health Security Implications

Ecuador's arboviral experience has important implications for global health security and pandemic preparedness beyond the Americas [173,174]:

Surveillance system resilience: The COVID-19 pandemic demonstrated the critical importance of maintaining essential health services during emergency responses. Countries should develop dual-capacity surveillance systems that can simultaneously respond to pandemic threats while monitoring endemic diseases.

Climate–health integration: The strong climate–epidemic associations documented in this study support integration of meteorological services with public health surveillance systems to enable climate-informed disease preparedness.

Multi-pathogen preparedness: Traditional pandemic preparedness focused on single emerging threats may be inadequate in settings where multiple pathogens circulate simultaneously. Preparedness strategies should consider complex pathogen interactions and their implications for surveillance, diagnosis, and control.

Equity and accessibility: The geographic and socioeconomic disparities in arboviral burden observed in Ecuador reflect broader patterns of health inequality that must be addressed through targeted interventions and resource allocation. The expansion of transmission zones poses risks not only through direct morbidity but also through secondary impacts on health system resilience, such as compromising the safety and availability of blood supplies in newly endemic regions [175].

5. Conclusions

This comprehensive 36-year analysis represents the most extensive temporal assessment of arboviral dynamics in Ecuador to date, revealing fundamental insights that extend far beyond national borders to inform global understanding of multi-pathogen transmission systems [2,3]. Our findings demonstrate that arboviral diseases function as an integrated epidemiological network rather than independent disease entities, with profound implications for surveillance, control, and pandemic preparedness strategies worldwide.

5.1. Principal Findings and Their Global Significance

Four key discoveries emerge from this analysis that fundamentally challenge conventional approaches to arboviral disease management:

First, the robust climate-epidemic associations ($r = 0.67$ for ENSO–dengue relationships) provide quantitative evidence for environmental forcing of transmission cycles, offering a 3–6 month predictive window that could transform epidemic preparedness from reactive to proactive approaches [12,26]. The identification of specific threshold effects (ONI > +0.8 °C associated with RR = 2.34) enables development of operational early warning systems with demonstrated 89% prediction accuracy, representing a paradigmatic shift toward climate-informed public health intervention.

Second, the paradoxical 27-fold reduction in severe dengue proportions (from 2.7% to 0.1%) following chikungunya and Zika emergence provides compelling evidence for immunological cross-protection between arboviruses [39,144]. This finding challenges the traditional antibody-dependent enhancement paradigm and suggests that controlled exposure to related arboviruses might provide population-level protection against severe disease manifestations, with profound implications for vaccine development and deployment strategies.

Third, the statistical evidence for competitive displacement between co-circulating arboviruses demonstrates complex viral interactions that influence transmission dynamics, epidemic timing, and disease severity patterns [9,140]. These multi-pathogen dynamics necessitate integrated surveillance and control approaches that consider viral interactions rather than treating each pathogen as an independent threat.

Fourth, the COVID-19 pandemic impact revealed both the vulnerability and resilience of arboviral surveillance systems, with a 73% reduction in case detection during lockdowns followed by explosive resurgence in 2024 [21,22]. This natural experiment demonstrates the critical importance of maintaining essential health services during pandemic responses and developing surveillance systems robust enough to function during healthcare system disruption.

5.2. Methodological Innovations and Analytical Contributions

This study introduces several methodological innovations that advance the field of infectious disease epidemiology [41,42]:

Diagnostic bias correction: Our approach to quantifying and correcting for evolving diagnostic capabilities over multi-decade time series provides a framework for analyzing long-term surveillance data with changing detection methods, applicable to other infectious diseases with evolving diagnostic landscapes.

Multi-pathogen interaction modeling: The integration of Vector Autoregression analysis with interrupted time series methods offers quantitative approaches for detecting and characterizing viral competition and displacement effects in real-world surveillance data.

Climate–epidemic integration: The combination of multiple climate indices with distributed lag non-linear models provides robust methods for identifying threshold effects and developing predictive early warning systems for climate-sensitive diseases.

Pandemic impact assessment: Our framework for quantifying COVID-19 effects on endemic disease surveillance offers methodological tools for assessing health system resilience and designing pandemic preparedness strategies that protect essential health services.

5.3. Implications for Global Arboviral Control and Pandemic Preparedness

5.3.1. Paradigm Shift Toward Integrated Management

The evidence for multi-pathogen interactions supports a fundamental paradigm shift from vertical, single-disease control programs to integrated arboviral management systems [156,157]. This transformation requires the following:

Unified surveillance platforms that simultaneously monitor all circulating arboviruses using standardized protocols, shared laboratory infrastructure, and coordinated response mechanisms, enabling detection of viral displacement patterns and cross-protective immunity effects.

Climate-informed preparedness based on quantitative early warning systems that trigger enhanced surveillance and preemptive intervention measures when environmental conditions indicate elevated epidemic risk, potentially preventing major outbreaks through proactive resource mobilization.

Flexible response capacity including diagnostic surge capability, point-of-care testing networks, and adaptable vector control strategies that can be rapidly scaled during multi-pathogen circulation periods or pandemic-related disruptions.

Community resilience building through decentralized surveillance systems, community health worker training, and social mobilization strategies that maintain disease monitoring and response capacity during healthcare system stress.

5.3.2. Innovation Priorities for the Next Decade

Our findings identify critical innovation priorities that could revolutionize arboviral disease management [2,169]:

Broad-spectrum diagnostics: Development of multiplex point-of-care tests capable of simultaneously detecting and differentiating multiple arboviruses with high sensitivity and specificity, addressing the diagnostic challenges revealed during the 2015–2017 triple epidemic.

Cross-protective vaccines: Investigation of multi-valent vaccination strategies that leverage immunological cross-protection between arboviruses, potentially providing broader population protection while minimizing adverse effects from antibody-dependent enhancement.

Predictive modeling systems: Integration of climate forecasting, epidemiological modeling, and artificial intelligence to develop operational early warning systems that provide quantitative epidemic risk assessments with sufficient lead time for preventive intervention.

Adaptive vector control: Development of climate-responsive and resistance-management strategies that can be rapidly deployed across diverse ecological settings and maintain effectiveness during multi-pathogen circulation periods.

5.4. Climate Change and Future Arboviral Risk

The strong climate–epidemic associations documented in this study have urgent implications for understanding future arboviral risk under accelerating climate change [11,168]. Key projections include the following:

Geographic expansion: Climate change will likely expand suitable arboviral habitat by 200–400 m elevation in Andean regions, potentially exposing an additional 2.8 million people to transmission risk by 2050, requiring preemptive surveillance and control infrastructure development.

Intensified transmission cycles: Stronger and more frequent El Niño events may amplify epidemic magnitude and frequency, necessitating enhanced preparedness capacity and more robust early warning systems to manage increased epidemic pressure.

Novel pathogen emergence: Changing climate patterns may facilitate the introduction and establishment of new arboviruses in previously unsuitable regions, highlighting the need for broad-spectrum surveillance and rapid response capabilities.

Vulnerable population protection: Indigenous communities, highland populations, and urban poor populations will face disproportionate risks from arboviral expansion, requiring targeted intervention strategies and equitable resource allocation.

5.5. Global Health Security and Pandemic Preparedness Lessons

Ecuador's arboviral experience provides critical lessons for global health security that extend beyond vector-borne diseases [173,174]:

Health system resilience: The COVID-19 pandemic demonstrated that effective pandemic preparedness requires maintaining essential health services while responding to novel threats, necessitating dual-capacity systems that can function under multiple simultaneous pressures.

Surveillance system robustness: Traditional surveillance approaches focused on single pathogens may be inadequate for managing complex multi-pathogen scenarios, requiring integrated monitoring systems that can detect and characterize pathogen interactions in real-time.

Early warning integration: Climate-informed early warning systems offer significant advantages over reactive surveillance approaches, providing actionable intelligence that enables proactive intervention before epidemic establishment.

Community engagement continuity: The collapse of community-based programs during COVID-19 lockdowns highlights the critical importance of maintaining social connections and community health capacity during emergency responses.

Equity and accessibility: Geographic and socioeconomic disparities in arboviral burden reflect broader patterns of health inequality that must be addressed through targeted interventions and resource allocation to achieve effective disease control.

5.6. Research Priorities and Future Directions

This analysis identifies several urgent research priorities that require immediate attention to advance arboviral science and improve control effectiveness [3,8]:

5.6.1. Immunological Mechanisms

Cross-protection validation: Prospective cohort studies with detailed immunological profiling are urgently needed to validate the multi-pathogen cross-protection hypothesis and identify biomarkers of protection that could guide vaccine development and deployment strategies.

T-cell cross-reactivity: Comprehensive characterization of cross-reactive T-cell responses between arboviruses could inform development of broad-spectrum therapeutic approaches and guide understanding of population immunity dynamics.

Severe disease mechanisms: Investigation of how sequential arboviral exposures influence immunopathology and disease severity could revolutionize clinical management and vaccine safety assessment.

5.6.2. Vector and Environmental Studies

Multi-pathogen vector competence: Laboratory and field studies examining within-mosquito viral interactions are essential for understanding displacement mechanisms and developing predictive models of multi-pathogen transmission.

Climate threshold validation: Experimental validation of climate-transmission thresholds identified in this study could improve predictive accuracy and enable development of location-specific early warning systems.

Ecosystem modeling: Integration of vector ecology, climate dynamics, and human behavior into comprehensive transmission models could enable more accurate risk assessment and intervention targeting.

5.6.3. Implementation Science

Integrated program evaluation: Comparative effectiveness research on integrated versus vertical disease control approaches is needed to optimize resource allocation and intervention strategies.

Community engagement innovation: Development and evaluation of culturally appropriate, technology-enabled community engagement strategies that maintain effectiveness during emergency responses.

Health system strengthening: Research on optimal designs for dual-capacity health systems that can simultaneously respond to pandemic threats while maintaining endemic disease control.

5.7. Call to Action: Transforming Arboviral Disease Management

The findings of this 36-year analysis demand immediate action to transform global approaches to arboviral disease management. The evidence for climate forcing, immunological cross-protection, and multi-pathogen interactions provides the scientific foundation for revolutionary changes in how we conceptualize, monitor, and control these diseases [2,176].

For policymakers: We urge policymakers to leverage the strong climate-epidemic links by investing immediately in integrated early warning systems and climate-informed health planning. Countries should prioritize development of dual-capacity surveillance systems that can maintain essential functions during pandemic responses while building resilience against climate-driven disease emergence.

For researchers: The immunological cross-protection hypothesis requires urgent validation through prospective studies that could fundamentally alter vaccination strategies and therapeutic approaches. Multi-pathogen interaction mechanisms demand investigation through interdisciplinary collaboration between virologists, entomologists, immunologists, and epidemiologists.

For public health practitioners: The evidence for viral displacement and competitive interactions necessitates immediate transition to integrated surveillance and control approaches that consider all circulating arboviruses simultaneously rather than focusing on individual pathogens.

For the global community: Climate change will continue to expand arboviral transmission risk, requiring international cooperation, technology transfer, and equitable resource allocation to protect vulnerable populations and prevent global disease emergence.

5.8. Final Reflections: Lessons from 36 Years of Arboviral Circulation

Ecuador's 36-year journey with arboviral diseases offers profound lessons that transcend national boundaries and disease categories. The country's experience demonstrates that infectious diseases are not isolated threats but components of complex adaptive systems shaped by climate, ecology, human behavior, and pathogen interactions over deep evolutionary time [134,173,177].

The COVID-19 pandemic provided an unprecedented natural experiment that revealed both the fragility and resilience of health systems, the interconnectedness of global disease threats, and the critical importance of maintaining essential services during emergency responses. The rapid rebound of dengue transmission following pandemic control measures serves as a stark reminder that endemic diseases remain present even when attention is focused elsewhere.

Perhaps most importantly, the evidence for immunological cross-protection between arboviruses suggests that nature may provide unexpected solutions to complex disease challenges. Understanding and harnessing these natural protective mechanisms could revolutionize our approach to vaccine development and disease prevention, offering hope for more effective and equitable control strategies.

As we face an uncertain future marked by accelerating climate change, increasing globalization, and emerging pandemic threats, the lessons learned from Ecuador's arboviral experience become increasingly relevant. Only through sustained investment in integrated surveillance systems, climate-informed preparedness strategies, innovative control approaches, and international cooperation can we hope to reduce the growing burden of vector-borne diseases while building resilience against future pandemic threats.

The path forward requires courage to abandon traditional vertical approaches in favor of innovative integrated strategies, wisdom to learn from nature's own protective mechanisms, and commitment to ensuring that the benefits of scientific advancement reach all populations, regardless of geography or economic status. Ecuador's arboviral story is ultimately a story of adaptation, resilience, and hope—qualities that will be essential as humanity confronts the disease challenges of the 21st century.

Critical Lessons for Pandemic Preparedness

- Climate-Epidemic Associations: Predictive early warning capabilities can transform reactive responses into proactive preparedness strategies.
- Multi-Pathogen Interactions: Complex epidemiological dynamics require integrated rather than vertical surveillance and control approaches.
- Diagnostic Surge Capacity: Broad-spectrum detection methods essential for emerging pathogen identification challenges.
- Resilient Surveillance Systems: Must maintain essential functions during healthcare system stress and pandemic responses.
- Cross-Protective Immunity: Related pathogens may provide unexpected population benefits for vaccine development and deployment strategies.

Supplementary Materials: The following supporting information can be downloaded at: https://www.mdpi.com/article/10.3390/v17091201/s1, Figure S1: Temporal Evolution of Arboviruses in Edcuador (1988–2024); Figure S2: Temporal Evolution of diseases in Edcuador (1988–2024); Figure S3: Comparospm of Arboviruses in Edcuador by Period (1988–2024); Figure S4: Cumulative Burden of

Arboviruses in Edcuador (1988–2024); Figure S5: Correlación entre Dengue, Zika y Chikungunya (2013–2023).

Author Contributions: Conceptualization, J.D.S. and C.Á.R.; methodology, J.D.S. and E.C.C.; software, E.C.C.; validation, J.A.S. and C.B.C.; formal analysis, J.D.S. and E.C.C.; investigation, all authors; data curation, E.C.C. and J.A.S.; writing—original draft preparation, J.D.S.; writing—review and editing, all authors; visualization, J.A.S. and E.C.C.; supervision, C.B.C.; project administration, J.D.S.; funding acquisition, J.D.S. and C.B.C. All authors have read and agreed to the published version of the manuscript.

Funding: This research was supported by Universidad Tecnológica Indoamérica, "Proyecto Semilla 2026". The article processing charge (APC) was funded by the institutional open access publication fund.

Institutional Review Board Statement: This study is based on publicly available, aggregated surveillance data that contains no individual identifiers. In accordance with Ecuadorian national regulations for the secondary use of public health data, formal ethical review and approval were not required for this type of study.

Informed Consent Statement: Not applicable. This study utilized aggregated surveillance data without individual identifiers.

Data Availability Statement: The original epidemiological datasets analyzed during this study are publicly available from Ecuador's Ministry of Public Health (MSP) at: https://www.salud.gob.ec/gacetas-epidemiologicas/ (accessed on 15 June 2024). Processed datasets used for statistical analysis, including climate–epidemic correlation matrices and diagnostic bias correction factors, are available from the corresponding author upon reasonable request. Climate data were obtained from NOAA Climate Prediction Center (https://origin.cpc.ncep.noaa.gov/products/analysis_monitoring/ensostuff/ONI_v5.php) (accessed on 15 June 2024) and are freely available. Population data were sourced from Ecuador's National Institute of Statistics (INEC) at: https://www.ecuadorencifras.gob.ec/ (accessed on 15 June 2024).

Acknowledgments: We extend our deepest gratitude to the Ministry of Public Health of Ecuador (MSP). We thank the healthcare workers throughout Ecuador who maintained surveillance activities during the challenging COVID-19 pandemic period, making this analysis possible. We acknowledge the computational resources provided by Universidad Tecnológica Indoamérica's High-Performance Computing Center.

Conflicts of Interest: The authors declare no conflict of interest.

Abbreviations

Abbreviation	**Definition**
ACF	Autocorrelation Function
ADE	Antibody-Dependent Enhancement
AIC	Akaike Information Criterion
CHIKV	Chikungunya virus
CI	Confidence Interval
COVID-19	Coronavirus Disease 2019
CUSUM	Cumulative Sum Control Chart
DENV	Dengue virus
DHF/DSS	Dengue Hemorrhagic Fever/Dengue Shock Syndrome
DLNM	Distributed Lag Non-linear Model
ELISA	Enzyme-Linked Immunosorbent Assay
ENSO	El Niño–Southern Oscillation
EW	Epidemiological Week
FFT	Fast Fourier Transform

GAM Generalized Additive Model
IgG Immunoglobulin G
IgM Immunoglobulin M
INAMHI Instituto Nacional de Meteorología e Hidrología
INEC Instituto Nacional de Estadística y Censos
INSPI Instituto Nacional de Investigación en Salud Pública
MEM Moving Epidemic Method
MSP Ministerio de Salud Pública
NDVI Normalized Difference Vegetation Index
NOAA National Oceanic and Atmospheric Administration
NS1 Non-structural protein 1
ONI Oceanic Niño Index
PAHO Pan American Health Organization
PCA Principal Component Analysis
PELT Pruned Exact Linear Time
RT-PCR Reverse Transcription-Polymerase Chain Reaction
SIVE Sistema de Vigilancia Epidemiológica
SST Sea Surface Temperature
STL Seasonal and Trend decomposition using Loess
VAR Vector Autoregression
WHO World Health Organization
ZIKV Zika virus

References

1. Messina, J.P.; Brady, O.J.; Golding, N.; Kraemer, M.U.G.; Wint, G.R.W.; Ray, S.E.; Pigott, D.M.; Shearer, F.M.; Johnson, K.; Earl, L.; et al. The current and future global distribution and population at risk of dengue. *Nat. Microbiol.* **2019**, *4*, 1508–1515. [CrossRef]
2. Brady, O.J.; Hay, S.I. The Rise and Fall of Dengue in the Americas. *Cell* **2019**, *178*, 774–776.
3. Wilder-Smith, A.; Gubler, D.J.; Weaver, S.C.; Monath, T.P.; Heymann, D.L.; Scott, T.W. Epidemic arboviral diseases: Priorities for research and public health. *Lancet Infect. Dis.* **2017**, *17*, e101–e106. [CrossRef] [PubMed]
4. Rodriguez-Morales, A.J.; Villamil-Gómez, W.E.; Franco-Paredes, C. The arboviral burden of disease caused by co-circulation and co-infection of dengue, chikungunya and Zika in the Americas. *Travel Med. Infect. Dis.* **2016**, *14*, 177–179. [CrossRef]
5. Powell, J.R.; Tabachnick, W.J. History of domestication and spread of *Aedes aegypti*—A review. *Mem. Inst. Oswaldo Cruz* **2013**, *108*, 11–17. [CrossRef] [PubMed]
6. Faria, N.R.; Azevedo, R.d.S.d.S.; Kraemer, M.U.G.; Souza, R.; Cunha, M.S.; Hill, S.C.; Thézé, J.; Bonsall, M.B.; Bowden, T.A.; Rissanen, I.; et al. Zika virus in the Americas: Early epidemiological and genetic findings. *Science* **2016**, *352*, 345–349. [CrossRef]
7. Weaver, S.C.; Lecuit, M. Chikungunya virus and the global spread of a mosquito-borne disease. *N. Engl. J. Med.* **2015**, *372*, 1231–1239. [CrossRef] [PubMed]
8. Vasilakis, N.; Weaver, S.C. The history and evolution of human arboviral disease. *PLoS Pathog.* **2017**, *13*, e1006092.
9. Vogels, C.B.F.; Rückert, C.; Cavany, S.M.; Perkins, T.A.; Ebel, G.D.; Grubaugh, N.D. Arbovirus coinfection and co-transmission: A neglected public health concern? *PLoS Biol.* **2019**, *17*, e3000130. [CrossRef] [PubMed]
10. Stewart-Ibarra, A.M.; Ryan, S.J.; Kenneson, A.; King, C.A.; Abbott, M.; Barbachano-Guerrero, A.; Beltrán-Ayala, E.; Borbor-Cordova, M.J.; Cárdenas, W.B.; Cueva, C.; et al. The Burden of Dengue Fever and Chikungunya in Southern Coastal Ecuador: Epidemiology, Clinical Presentation, and Phylogenetics from the First Two Years of a Prospective Study. *Am. J. Trop. Med. Hyg.* **2018**, *98*, 1444–1459. [CrossRef] [PubMed]
11. Ryan, S.J.; Carlson, C.J.; Mordecai, E.A.; Johnson, L.R. Global expansion and redistribution of Aedes-borne virus transmission risk with climate change. *PLoS Negl. Trop. Dis.* **2019**, *13*, e0007213. [CrossRef] [PubMed]
12. Lowe, R.; Stewart-Ibarra, A.M.; Petrova, D.; García-Díez, M.; Borbor-Cordova, M.J.; Mejía, R.; Regato, M.; Rodó, X. Climate services for health: Predicting the evolution of the 2016 dengue season in Machala, Ecuador. *Lancet Planet. Health* **2017**, *1*, e142–e151. [CrossRef] [PubMed]
13. Sippy, R.; Herrera, D.; Gaus, D.; Gangnon, R.E.; Patz, J.A.; Osorio, J.E. Seasonal patterns of dengue fever in rural Ecuador: 2009–2016. *PLoS Negl. Trop. Dis.* **2019**, *13*, e0007360. [CrossRef] [PubMed]
14. Katzelnick, L.C.; Quentin, E.; Colston, S.; Ha, T.A.; Andrade, P.; Eisenberg, J.N.S.; Ponce, P.; Coloma, J.; Cevallos, V. Increasing transmission of dengue virus across ecologically diverse regions of Ecuador and associated risk factors. *PLoS Negl. Trop. Dis.* **2024**, *18*, e0011408. [CrossRef] [PubMed]

15. Coloma, J.; Harris, E. Molecular and virological study of dengue in Ecuador. *Rev. Panam. Salud Publica* **2010**, *27*, 398–402.
16. Stewart-Ibarra, A.M.; Lowe, R. Climate and non-climate drivers of dengue epidemics in southern coastal Ecuador. *Am. J. Trop. Med. Hyg.* **2013**, *88*, 971–981. [CrossRef] [PubMed]
17. Pan American Health Organization. *Epidemiological Update: Dengue*; PAHO/WHO: Washington, DC, USA, 2019. Available online: https://iris.paho.org/handle/10665.2/51496 (accessed on 26 July 2025).
18. Márquez, S.; Lee, G.O.; Andrade, P.; Zuniga, J.; Trueba, G.; Eisenberg, J.N.S.; Coloma, J. A Chikungunya Outbreak in a Dengue-endemic Region in Rural Northern Coastal Ecuador. *Am. J. Trop. Med. Hyg.* **2022**, *107*, 1226–1233. [CrossRef]
19. Waggoner, J.J.; Pinsky, B.A. Zika virus: Diagnostics for an emerging pandemic threat. *J. Clin. Microbiol.* **2016**, *54*, 860–867. [CrossRef] [PubMed]
20. Musso, D.; Gubler, D.J. Zika virus. *Clin. Microbiol. Rev.* **2016**, *29*, 487–524. [CrossRef]
21. Cardona-Ospina, J.A.; Arteaga-Livias, K.; Villamil-Gómez, W.E.; Pérez-Díaz, C.E.; Bonilla-Aldana, D.K.; Mondragon-Cardona, A.; Rodriguez-Morales, A.J.; Solarte-Portilla, M.; Martinez, E.; Millan-Oñate, J.; et al. Dengue and COVID-19, overlapping epidemics? An analysis from Colombia. *J. Med. Virol.* **2021**, *93*, 522–527. [CrossRef]
22. Quintero, B.; Ramón-Ochoa, A.X.; Morillo-Puente, S.; Tenezaca-Ramón, D.A.; Cevallos-Naranjo, A.S. Emergence and expansion of dengue in Paltas: Possible implications of the COVID-19 pandemic and climatic variations. *Trop. Med. Health* **2025**, *53*, 12. [CrossRef] [PubMed]
23. Lorenz, C.; Azevedo, T.S.; Chiaravalloti-Neto, F. COVID-19 and dengue fever: A dangerous combination for the health system in Brazil. *Travel Med. Infect. Dis.* **2020**, *35*, 101659. [CrossRef]
24. Garcia-Bereguiain, M.A.; Gonzalez, E.; Vivanco, E.; Vega-Mariño, P.; Romero, C.; Comas-García, A.; Vera, E.; Loaiza, O.; Bonilla, F.; Aguirre, M.; et al. COVID-19 in Ecuador: Preliminary molecular epidemiological analysis and public health implications. *J. Glob. Health* **2020**, *10*, 020354. [CrossRef]
25. Kraemer, M.U.G.; Sinka, M.E.; Duda, K.A.; Mylne, A.Q.N.; Shearer, F.M.; Barker, C.M.; Moore, C.G.; Carvalho, R.G.; Coelho, G.E.; Van Bortel, W.; et al. The global distribution of the arbovirus vectors Aedes aegypti and Ae. albopictus. *eLife* **2015**, *4*, e08347. [CrossRef] [PubMed]
26. Mordecai, E.A.; Caldwell, J.M.; Grossman, M.K.; Lippi, C.A.; Johnson, L.R.; Neira, M.; Rohr, J.R.; Ryan, S.J.; Savage, V.; Shocket, M.S.; et al. Thermal biology of mosquito-borne disease. *Ecol. Lett.* **2019**, *22*, 1690–1708. [CrossRef] [PubMed]
27. Freire-Paspuel, B.; Vega-Mariño, P.; Velez, A.; Cruz, M.; Garcia-Bereguiain, M.A. Sample pooling of RNA extracts to speed up SARS-CoV-2 diagnosis using CDC FDA EUA RT-qPCR kit. *Virus Res.* **2020**, *290*, 198173. [CrossRef] [PubMed]
28. Collins, M.H.; McGowan, E.; Jadi, R.; Young, E.; Lopez, C.A.; Baric, R.S.; Lazear, H.M.; de Silva, A.M. Lack of Durable Cross-Neutralizing Antibodies Against Zika Virus from Dengue Virus Infection. *Emerg. Infect. Dis.* **2017**, *23*, 773–781. [CrossRef] [PubMed]
29. Perera, R.A.P.M.; Tso, E.; Tsang, O.T.Y.; Tsang, D.N.C.; Fung, K.; Leung, Y.W.Y.; Chin, A.W.H.; Chu, D.K.W.; Cheng, S.M.S.; Poon, L.L.M.; et al. SARS-CoV-2 Virus Culture and Subgenomic RNA for Respiratory Specimens from Patients with Mild Coronavirus Disease. *Emerg. Infect. Dis.* **2020**, *26*, 2701–2704. [CrossRef]
30. Grubaugh, N.D.; Saraf, S.; Gangavarapu, K.; Watts, A.; Tan, A.L.; Oidtman, R.J.; Ladner, J.T.; Oliveira, G.; Matteson, N.L.; Kraemer, M.U.G.; et al. Travel Surveillance and Genomics Uncover a Hidden Zika Outbreak during the Waning Epidemic. *Cell* **2019**, *178*, 1057–1071.e11. [CrossRef] [PubMed]
31. Duffy, M.R.; Chen, T.H.; Hancock, W.T.; Powers, A.M.; Kool, J.L.; Lanciotti, R.S.; Pretrick, M.; Marfel, M.; Holzbauer, S.; Dubray, C.; et al. Zika virus outbreak on Yap Island, Federated States of Micronesia. *N. Engl. J. Med.* **2009**, *360*, 2536–2543. [CrossRef] [PubMed]
32. Katzelnick, L.C.; Montoya, M.; Gresh, L.; Balmaseda, A.; Harris, E. Neutralizing antibody titers against dengue virus correlate with protection from symptomatic infection in a longitudinal cohort. *Proc. Natl. Acad. Sci. USA* **2016**, *113*, 728–733. [CrossRef] [PubMed]
33. Salje, H.; Cummings, D.A.T.; Rodriguez-Barraquer, I.; Katzelnick, L.C.; Lessler, J.; Klungthong, C.; Thaisomboonsuk, B.; Nisalak, A.; Weg, A.; Ellison, D.; et al. Reconstruction of antibody dynamics and infection histories to evaluate dengue risk. *Nature* **2018**, *557*, 719–723. [CrossRef] [PubMed]
34. Rodriguez-Barraquer, I.; Costa, F.; Nascimento, E.J.M.; Nery, N., Jr.; Castanha, P.M.S.; Sacramento, G.A.; Cruz, J.; Carvalho, M.; De Olivera, D.; Hagan, J.E.; et al. Impact of preexisting dengue immunity on Zika virus emergence in a dengue endemic region. *Science* **2019**, *363*, 607–610. [CrossRef] [PubMed]
35. Hyndman, R.J.; Athanasopoulos, G. *Forecasting: Principles and Practice*, 2nd ed.; OTexts: Melbourne, Australia, 2018.
36. Killick, R.; Fearnhead, P.; Eckley, I.A. Optimal detection of changepoints with a linear computational cost. *J. Am. Stat. Assoc.* **2012**, *107*, 1590–1598. [CrossRef]
37. Lowe, R.; Bailey, T.C.; Stephenson, D.B.; Graham, R.J.; Coelho, C.A.S.; Carvalho, M.S.; Barcellos, C. Spatio-temporal modelling of climate-sensitive disease risk: Towards an early warning system for dengue in Brazil. *Comput. Geosci.* **2011**, *37*, 371–381. [CrossRef]

38. Tamerius, J.; Nelson, M.I.; Zhou, S.Z.; Viboud, C.; Miller, M.A.; Alonso, W.J. Global influenza seasonality: Reconciling patterns across temperate and tropical regions. *Environ. Health Perspect.* **2011**, *119*, 439–445. [CrossRef] [PubMed]
39. Andrade, P.; Gimblet-Ochieng, C.; Modirian, F.; Collins, M.; Cardenas, M.; Katzelnick, L.C.; Montoya, M.; Michlmayr, D.; Kuan, G.; Balmaseda, A.; et al. Impact of pre-existing dengue immunity on human antibody and memory B cell responses to Zika. *Nat. Commun.* **2019**, *10*, 938. [CrossRef] [PubMed]
40. Andrade, P.; Narvekar, P.; Montoya, M.; Michlmayr, D.; Balmaseda, A.; Coloma, J.; Harris, E. Primary and Secondary Dengue Virus Infections Elicit Similar Memory B-Cell Responses, but Breadth to Other Serotypes and Cross-Reactivity to Zika Virus Is Higher in Secondary Dengue. *J. Infect. Dis.* **2020**, *222*, 590–600. [CrossRef]
41. Rothman, K.J.; Greenland, S.; Lash, T.L. *Modern Epidemiology*, 3rd ed.; Lippincott Williams & Wilkins: Philadelphia, PA, USA, 2008; p. 758.
42. Szklo, M.; Nieto, F.J. *Epidemiology: Beyond the Basics*, 3rd ed.; Jones & Bartlett Learning: Burlington, MA, USA, 2014; p. 516.
43. Bernal, J.L.; Cummins, S.; Gasparrini, A. Interrupted time series regression for the evaluation of public health interventions: A tutorial. *Int. J. Epidemiol.* **2017**, *46*, 348–355. [CrossRef]
44. Kontopantelis, E.; Doran, T.; Springate, D.A.; Buchan, I.; Reeves, D. Regression based quasi-experimental approach when randomisation is not possible: Interrupted time series analysis. *BMJ* **2015**, *350*, h2750. [CrossRef] [PubMed]
45. Instituto Nacional de Estadística y Censos del Ecuador. *Proyecciones Poblacionales*; INEC: Quito, Ecuador, 2023. Available online: https://www.ecuadorencifras.gob.ec/proyecciones-poblacionales/ (accessed on 26 July 2025).
46. Young, K.R.; Ulloa, C.U.; Luteyn, J.L.; Knapp, S. Plant evolution and endemism in Andean South America: An introduction. *Bot. Rev.* **2002**, *68*, 4–21. [CrossRef]
47. von Elm, E.; Altman, D.G.; Egger, M.; Pocock, S.J.; Gøtzsche, P.C.; Vandenbroucke, J.P. The Strengthening the Reporting of Observational Studies in Epidemiology (STROBE) statement: Guidelines for reporting observational studies. *Lancet* **2007**, *370*, 1453–1457. [CrossRef] [PubMed]
48. Thacker, S.B.; Qualters, J.R.; Lee, L.M. Public health surveillance in the United States: Evolution and challenges. *MMWR Suppl.* **2012**, *61*, 3–9. [PubMed]
49. German, R.R.; Lee, L.M.; Horan, J.M.; Milstein, R.L.; Pertowski, C.A.; Waller, M.N.; Guidelines Working Group. Updated guidelines for evaluating public health surveillance systems: Recommendations from the guidelines working group. *MMWR Recomm. Rep.* **2001**, *50*, 1–35. [PubMed]
50. Ministerio de Salud Pública del Ecuador. *Protocolo de Vigilancia Epidemiológica para Arbovirosis*; MSP: Quito, Ecuador, 2023. Available online: https://www.salud.gob.ec/wp-content/uploads/2023/02/Gaceta-SE-3_2023.pdf (accessed on 26 July 2025).
51. World Health Organization. *Dengue: Guidelines for Diagnosis, Treatment, Prevention and Control*; WHO Press: Geneva, Switzerland, 2009.
52. Pan American Health Organization. *Tool for the Diagnosis and Care of Patients with Suspected Arboviral Diseases*; PAHO: Washington, DC, USA, 2017.
53. Guzman, M.G.; Halstead, S.B.; Artsob, H.; Buchy, P.; Farrar, J.; Gubler, D.J.; Hunsperger, E.; Kroeger, A.; Margolis, H.S.; Martínez, E.; et al. Dengue: A continuing global threat. *Nat. Rev. Microbiol.* **2010**, *8*, S7–S16. [CrossRef] [PubMed]
54. Peeling, R.W.; Artsob, H.; Pelegrino, J.L.; Buchy, P.; Cardosa, M.J.; Devi, S.; Enria, D.A.; Farrar, J.; Gubler, D.J.; Guzman, M.G.; et al. Evaluation of diagnostic tests: Dengue. *Nat. Rev. Microbiol.* **2010**, *8*, S30–S38. [CrossRef] [PubMed]
55. Blacksell, S.D. Commercial dengue rapid diagnostic tests for point-of-care application: Recent evaluations and future needs? *J. Biomed. Biotechnol.* **2012**, *2012*, 151967. [CrossRef]
56. Hunsperger, E.A.; Yoksan, S.; Buchy, P.; Nguyen, V.C.; Sekaran, S.D.; Enria, D.A.; Vazquez, S.; Cartozian, E.; Pelegrino, J.L.; Artsob, H.; et al. Evaluation of commercially available anti-dengue virus immunoglobulin M tests. *Emerg. Infect. Dis.* **2009**, *15*, 436–440. [CrossRef]
57. Lima, M.d.R.Q.; Nogueira, R.M.R.; Schatzmayr, H.G.; dos Santos, F.B. Comparison of three commercially available dengue NS1 antigen capture assays for acute diagnosis of dengue in Brazil. *PLoS Negl. Trop. Dis.* **2010**, *4*, e738. [CrossRef]
58. Santiago, G.A.; Vergne, E.; Quiles, Y.; Cosme, J.; Vazquez, J.; Medina, J.F.; Medina, F.; Colón, C.; Margolis, H.; Muñoz-Jordán, J.L. Analytical and clinical performance of the CDC real time RT-PCR assay for detection and typing of dengue virus. *PLoS Negl. Trop. Dis.* **2013**, *7*, e2311. [CrossRef]
59. Lanciotti, R.S.; Kosoy, O.L.; Laven, J.J.; Velez, J.O.; Lambert, A.J.; Johnson, A.J.; Stanfield, S.M.; Duffy, M.R. Genetic and serologic properties of Zika virus associated with an epidemic, Yap State, Micronesia, 2007. *Emerg. Infect. Dis.* **2008**, *14*, 1232–1239. [CrossRef]
60. Domingo, C.; Palacios, G.; Jabado, O.; Reyes, N.; Niedrig, M.; Gascón, J.; Cabrerizo, M.; Lipkin, W.I.; Tenorio, A. Use of a short fragment of the C-terminal E gene for detection and characterization of two new lineages of dengue virus 1 in India. *J. Clin. Microbiol.* **2006**, *44*, 1519–1529. [CrossRef]

61. Harris, E.; Videa, E.; Pérez, L.; Sandoval, E.; Téllez, Y.; Pérez, M.L.; Cuadra, R.; Rocha, J.; Idiaquez, W.; Alonso, R.E.; et al. Clinical, epidemiologic, and virologic features of dengue in the 1998 epidemic in Nicaragua. *Am. J. Trop. Med. Hyg.* **2000**, *63*, 5–11. [CrossRef]
62. Mansfield, K.L.; Horton, D.L.; Johnson, N.; Li, L.; Barrett, A.D.; Smith, D.J.; Galbraith, S.E.; Solomon, T.; Fooks, A.R. Flavivirus-induced antibody cross-reactivity. *J. Gen. Virol.* **2011**, *92*, 2821–2829. [CrossRef] [PubMed]
63. Felix, A.C.; Romano, C.M.; Centrone, C.d.C.; Rodrigues, C.L.; Villas-Boas, L.; Araújo, E.S.; de Matos, A.M.; Carvalho, K.I.; Martelli, C.M.; Kallas, E.G.; et al. Low sensitivity of NS1 protein tests evidenced during a dengue type 2 virus outbreak in Santos, Brazil, in 2010. Clin. Vaccine Immunol. 2012, 19, 1972–1976. https://doi.org/10.1128/CVI.00535-12[CrossRef] [PubMed]
64. National Oceanic and Atmospheric Administration. Oceanic Niño Index (ONI). Available online: https://origin.cpc.ncep.noaa.gov/products/analysis_monitoring/ensostuff/ONI_v5.php (accessed on 26 July 2025).
65. Trenberth, K.E. The definition of El Niño. *Bull. Am. Meteorol. Soc.* **1997**, *78*, 2771–2777. [CrossRef]
66. Instituto Nacional de Meteorología e Hidrología del Ecuador. *Boletín Climatológico Nacional 2023*; INAMHI: Quito, Ecuador, 2024. Available online: https://servicios.inamhi.gob.ec/clima/ (accessed on 26 July 2025).
67. Peterson, T.C.; Vose, R.S. An overview of the Global Historical Climatology Network temperature database. *Bull. Am. Meteorol. Soc.* **1997**, *78*, 2837–2849. [CrossRef]
68. Tucker, C.J. Red and photographic infrared linear combinations for monitoring vegetation. *Remote Sens. Environ.* **1979**, *8*, 127–150. [CrossRef]
69. Anyamba, A.; Chretien, J.P.; Small, J.; Tucker, C.J.; Formenty, P.B.; Richardson, J.H.; Britch, S.C.; Schnabel, D.C.; Erickson, R.L.; Linthicum, K.J. Prediction of a Rift Valley fever outbreak. *Proc. Natl. Acad. Sci. USA* **2009**, *106*, 955–959. [CrossRef] [PubMed]
70. Huang, B.; Thorne, P.W.; Banzon, V.F.; Boyer, T.; Chepurin, G.; Lawrimore, J.H.; Menne, M.J.; Smith, T.M.; Vose, R.S.; Zhang, H.M. Extended Reconstructed Sea Surface Temperature, Version 5 (ERSSTv5): Upgrades, Validations, and Intercomparisons. *J. Clim.* **2017**, *30*, 8179–8205. [CrossRef]
71. Reynolds, R.W.; Smith, T.M.; Liu, C.; Chelton, D.B.; Casey, K.S.; Schlax, M.G. Daily high-resolution-blended analyses for sea surface temperature. *J. Clim.* **2007**, *20*, 5473–5496. [CrossRef]
72. Hogan, A.B.; Jewell, B.L.; Sherrard-Smith, E.; Vesga, J.F.; Watson, O.J.; Whittaker, C.; Hamlet, A.; Smith, J.A.; Winskill, P.; Verity, R.; et al. Potential impact of the COVID-19 pandemic on HIV, tuberculosis, and malaria in low-income and middle-income countries: A modelling study. *Lancet Glob. Health* **2020**, *8*, e1132–e1141. [CrossRef] [PubMed]
73. Hale, T.; Angrist, N.; Goldszmidt, R.; Kira, B.; Petherick, A.; Phillips, T.; Webster, S.; Cameron-Blake, E.; Hallas, L.; Majumdar, S.; et al. A global panel database of pandemic policies (Oxford COVID-19 Government Response Tracker). *Nat. Hum. Behav.* **2021**, *5*, 529–538. [CrossRef]
74. Ministerio de Salud Pública del Ecuador. *Informe de Gestión COVID-19: Impacto en el Sistema de Salud*; MSP: Quito, Ecuador, 2021.
75. Litewka, S.G.; Heitman, E. Latin American healthcare systems in times of pandemic. *Dev. World Bioeth.* **2020**, *20*, 69–73. [CrossRef] [PubMed]
76. Buehler, J.W.; Hopkins, R.S.; Overhage, J.M.; Sosin, D.M.; Tong, V.; CDC Working Group. Framework for evaluating public health surveillance systems for early detection of outbreaks: Recommendations from the CDC working group. *MMWR Recomm. Rep.* **2004**, *53*, 1–11. [PubMed]
77. World Health Organization. *Global Strategy for Dengue Prevention and Control 2012–2020*; WHO Press: Geneva, Switzerland, 2012.
78. Morrison, A.C.; Zielinski-Gutierrez, E.; Scott, T.W.; Rosenberg, R. Defining challenges and proposing solutions for control of the virus vector Aedes aegypti. *PLoS Med.* **2008**, *5*, e68. [CrossRef] [PubMed]
79. Farrington, C.P.; Andrews, N.J.; Beale, A.D.; Catchpole, M.A. A statistical algorithm for the early detection of outbreaks of infectious disease. *J. R. Stat. Soc. Ser. A* **1996**, *159*, 547–563. [CrossRef]
80. Noufaily, A.; Enki, D.G.; Farrington, P.; Garthwaite, P.; Andrews, N.; Charlett, A. An improved algorithm for outbreak detection in multiple surveillance systems. *Stat. Med.* **2013**, *32*, 1206–1222. [CrossRef]
81. Aminikhanghahi, S.; Cook, D.J. A survey of methods for time series change point detection. *Knowl. Inf. Syst.* **2017**, *51*, 339–367. [CrossRef] [PubMed]
82. Zeileis, A.; Leisch, F.; Hornik, K.; Kleiber, C. strucchange: An R package for testing for structural change in linear regression models. *J. Stat. Softw.* **2002**, *7*, 1–38. [CrossRef]
83. Doyle, T.J.; Glynn, M.K.; Groseclose, S.L. Completeness of notifiable infectious disease reporting in the United States: An analytical literature review. *Am. J. Epidemiol.* **2002**, *155*, 866–874. [CrossRef]
84. Brookmeyer, R.; Stroup, D.F. *Monitoring the Health of Populations: Statistical Principles and Methods for Public Health Surveillance*; Oxford University Press: New York, NY, USA, 2004.
85. Cleveland, R.B.; Cleveland, W.S.; McRae, J.E.; Terpenning, I. STL: A seasonal-trend decomposition procedure based on loess. *J. Off. Stat.* **1990**, *6*, 3–73.
86. Gasparrini, A. Modeling exposure-lag-response associations with distributed lag non-linear models. *Stat. Med.* **2014**, *33*, 881–899. [CrossRef] [PubMed]

87. Armstrong, B. Models for the relationship between ambient temperature and daily mortality. *Epidemiology* **2006**, *17*, 624–631 [CrossRef] [PubMed]
88. Chatfield, C. *The Analysis of Time Series: An Introduction*, 6th ed.; Chapman and Hall/CRC: Boca Raton, FL, USA, 2003.
89. Shumway, R.H.; Stoffer, D.S. *Time Series Analysis and Its Applications: With R Examples*, 4th ed.; Springer: New York, NY, USA, 2017
90. Gasparrini, A.; Armstrong, B.; Kenward, M.G. Distributed lag non-linear models. *Stat. Med.* **2010**, *29*, 2224–2234. [CrossRef]
91. Gasparrini, A.; Guo, Y.; Sera, F.; Vicedo-Cabrera, A.M.; Huber, V.; Tong, S.; de Sousa Zanotti Stagliorio Coelho, M.; Nascimento Saldiva, P.H.; Lavigne, E.; Matus Correa, P.; et al. Projections of temperature-related excess mortality under climate change scenarios. *Lancet Planet. Health* **2017**, *1*, e360–e367. [CrossRef] [PubMed]
92. Breiman, L.; Friedman, J.; Stone, C.J.; Olshen, R.A. *Classification and Regression Trees*; CRC Press: Boca Raton, FL, USA, 1984.
93. Therneau, T.; Atkinson, B. *rpart: Recursive Partitioning and Regression Trees*. R Package Version 4.1-19, 2023. Available online https://CRAN.R-project.org/package=rpart (accessed on 26 July 2025).
94. Hamilton, J.D. *Time Series Analysis*; Princeton University Press: Princeton, NJ, USA, 1994.
95. Earn, D.J.D.; Rohani, P.; Bolker, B.M.; Grenfell, B.T. A simple model for complex dynamical transitions in epidemics. *Science* **2000** *287*, 667–670. [CrossRef] [PubMed]
96. Pfaff, B. VAR, SVAR and SVEC models: Implementation within R package vars. *J. Stat. Softw.* **2008**, *27*, 1–32. [CrossRef]
97. Lütkepohl, H. *New Introduction to Multiple Time Series Analysis*; Springer: Berlin, Germany, 2005.
98. Granger, C.W.J. Investigating causal relations by econometric models and cross-spectral methods. *Econometrica* **1969**, *37*, 424–438 [CrossRef]
99. Toda, H.Y.; Yamamoto, T. Statistical inference in vector autoregressions with possibly integrated processes. *J. Econom.* **1995**, *66* 225–250. [CrossRef]
100. Jolliffe, I.T.; Cadima, J. Principal component analysis: A review and recent developments. *Philos. Trans. R. Soc. A* **2016**, *374* 20150202. [CrossRef] [PubMed]
101. Abdi, H.; Williams, L.J. Principal component analysis. *Wiley Interdiscip. Rev. Comput. Stat.* **2010**, *2*, 433–459. [CrossRef]
102. Penfold, R.B.; Zhang, F. Use of interrupted time series analysis in evaluating health care quality improvements. *Acad. Pediatr* **2013**, *13*, S38–S44. [CrossRef] [PubMed]
103. Kelsey, J.L.; Whittemore, A.S.; Evans, A.S.; Thompson, W.D. *Methods in Observational Epidemiology*, 2nd ed.; Oxford University Press: New York, NY, USA, 1996.
104. Pan American Health Organization. PLISA Health Information Platform for the Americas. Available online: https://www3.paho.org/data/index.php/en/ (accessed on 26 July 2025).
105. World Health Organization. Global Health Observatory Data Repository. Available online: https://apps.who.int/gho/data/ (accessed on 26 July 2025).
106. van den Broeck, J.; Cunningham, S.A.; Eeckels, R.; Herbst, K. Data cleaning: Detecting, diagnosing, and editing data abnormalities *PLoS Med.* **2005**, *2*, e267. [CrossRef]
107. English, D.R.; Daya, S. Data quality in health research. In *Encyclopedia of Health and Behavior*; Anderson, N.B., Ed.; Sage Publications: Thousand Oaks, CA, USA, 2004; pp. 234–237.
108. Rubin, D.B. *Multiple Imputation for Nonresponse in Surveys*; John Wiley & Sons: New York, NY, USA, 1987.
109. White, I.R.; Royston, P.; Wood, A.M. Multiple imputation using chained equations: Issues and guidance for practice. *Stat. Med* **2011**, *30*, 377–399. [CrossRef] [PubMed]
110. Hsu, C.C.; Sandford, B.A. The Delphi technique: Making sense of consensus. *Pract. Assess. Res. Eval.* **2007**, *12*, 10. [CrossRef]
111. Diamond, I.R.; Grant, R.C.; Feldman, B.M.; Pencharz, P.B.; Ling, S.C.; Moore, A.M.; Wales, P.W. Defining consensus: A systematic review recommends methodologic criteria for reporting of Delphi studies. *J. Clin. Epidemiol.* **2014**, *67*, 401–409. [CrossRef] [PubMed]
112. Wickham, H.; Averick, M.; Bryan, J.; Chang, W.; McGowan, L.D.; François, R.; Grolemund, G.; Hayes, A.; Henry, L.; Hester J.; et al. Welcome to the tidyverse. *J. Open Source Softw.* **2019**, *4*, 1686. [CrossRef]
113. Hyndman, R.J.; Khandakar, Y. Automatic time series forecasting: The forecast package for R. *J. Stat. Softw.* **2008**, *27*, 1–2 [CrossRef]
114. Killick, R.; Eckley, I.A. changepoint: An R Package for Changepoint Analysis. *J. Stat. Softw.* **2014**, *58*, 1–19. [CrossRef]
115. Gasparrini, A. Distributed lag linear and non-linear models in R: The package dlnm. *J. Stat. Softw.* **2011**, *43*, 1–20. [CrossRef] [PubMed]
116. Hill, A.B. The environment and disease: Association or causation? *Proc. R. Soc. Med.* **1965**, *58*, 295–300. [CrossRef] [PubMed]
117. Lessler, J.; Chaisson, L.H.; Kucirka, L.M.; Bi, Q.; Grantz, K.; Salje, H.; Carcelen, A.C.; Ott, C.T.; Sheffield, J.S.; Ferguson, N.M.; et a Assessing the global threat from Zika virus. *Science* **2016**, *353*, aaf8160. [CrossRef] [PubMed]
118. Robinson, W.S. Ecological correlations and the behavior of individuals. *Am. Sociol. Rev.* **1950**, *15*, 351–357. [CrossRef]
119. Morgenstern, H. Ecologic studies in epidemiology: Concepts, principles, and methods. *Annu. Rev. Public Health* **1995**, *16*, 61–8 [CrossRef] [PubMed]

120. Council for International Organizations of Medical Sciences. *International Ethical Guidelines for Health-Related Research Involving Humans*, 4th ed.; CIOMS: Geneva, Switzerland, 2016.
121. World Medical Association. World Medical Association Declaration of Helsinki: Ethical principles for medical research involving human subjects. *JAMA* **2013**, *310*, 2191–2194. [CrossRef] [PubMed]
122. *ISO/IEC 27001:2013*; Information Technology—Security Techniques—Information Security Management Systems—Requirements. ISO: Geneva, Switzerland, 2013.
123. Halstead, S.B. Dengvaxia sensitizes seronegatives to vaccine enhanced disease regardless of age. *Vaccine* **2017**, *35*, 6355–6358. [CrossRef]
124. Erdman, C.; Emerson, J.W. bcp: An R package for performing a Bayesian analysis of change point problems. *J. Stat. Softw.* **2007**, *23*, 1–13. [CrossRef]
125. Anderson, R.M.; May, R.M. *Infectious Diseases of Humans: Dynamics and Control*; Oxford University Press: Oxford, UK, 1991.
126. Keeling, M.J.; Rohani, P. *Modeling Infectious Diseases in Humans and Animals*; Princeton University Press: Princeton, NJ, USA, 2008.
127. Vega-Corredor, M.C.; Tiga-Loza, A.; Torres, A.L.; Morales-Betancourt, M.A.; López-Corzo, J.C. Spatial and temporal analysis of dengue transmission patterns in tropical Andean settings: A case study from Colombia. *Rev. Salud Publica* **2014**, *16*, 845–858.
128. Colón-González, F.J.; Fezzi, C.; Lake, I.R.; Hunter, P.R. The effects of weather and climate change on dengue. *PLoS Negl. Trop. Dis.* **2013**, *7*, e2503. [CrossRef]
129. Tesla, B.; Demakovsky, L.R.; Mordecai, E.A.; Ryan, S.J.; Bonds, M.H.; Ngonghala, C.N.; Brindley, M.A.; Murdock, C.C. Temperature drives Zika virus transmission: Evidence from empirical and mathematical models. *Proc. R. Soc. B* **2018**, *285*, 20180795. [CrossRef] [PubMed]
130. Lambrechts, L.; Paaijmans, K.P.; Fansiri, T.; Carrington, L.B.; Kramer, L.D.; Thomas, M.B.; Scott, T.W. Impact of daily temperature fluctuations on dengue virus transmission by Aedes aegypti. *Proc. Natl. Acad. Sci. USA* **2011**, *108*, 7460–7465. [CrossRef] [PubMed]
131. Carrington, L.B.; Seifert, S.N.; Willits, N.H.; Lambrechts, L.; Scott, T.W. Large diurnal temperature fluctuations negatively influence Aedes aegypti (Diptera: Culicidae) life-history traits. *J. Med. Entomol.* **2013**, *50*, 43–51. [CrossRef] [PubMed]
132. Naish, S.; Dale, P.; Mackenzie, J.S.; McBride, J.; Mengersen, K.; Tong, S. Climate change and dengue: A critical and systematic review of quantitative modelling approaches. *BMC Infect. Dis.* **2014**, *14*, 167. [CrossRef]
133. Vezzulli, L.; Grande, C.; Reid, P.C.; Hélaouët, P.; Edwards, M.; Höfle, M.G.; Brettar, I.; Colwell, R.R.; Pruzzo, C. Climate influence on Vibrio and associated human diseases during the past half-century in the coastal North Atlantic. *Proc. Natl. Acad. Sci. USA* **2016**, *113*, E5062–E5071. [CrossRef] [PubMed]
134. McMichael, A.J.; Woodruff, R.E.; Hales, S. Climate change and human health: Present and future risks. *Lancet* **2006**, *367*, 859–869. [CrossRef] [PubMed]
135. Confalonieri, U.; Menne, B.; Akhtar, R.; Ebi, K.L.; Hauengue, M.; Kovats, R.S.; Revich, B.; Woodward, A. Human health. In *Climate Change 2007: Impacts, Adaptation and Vulnerability. Contribution of Working Group II to the Fourth Assessment Report of the Intergovernmental Panel on Climate Change*; Parry, M.L., Canziani, O.F., Palutikof, J.P., van der Linden, P.J., Hanson, C.E., Eds.; Cambridge University Press: Cambridge, UK, 2007; pp. 391–431.
136. Rehman, A.U.; Hassanein, E.; Sohail, S.S.; Alomrani, A.; Aljubran, F.; Alsayed, A.R.; Alsamiri, H.M.; Ahmad, R.; Khan, N.A. Predictive modeling of the El Niño-Southern Oscillation effect on dengue transmission in Ecuador. *PLoS ONE* **2022**, *17*, e0275745. [CrossRef]
137. Patterson, J.; Sammon, M.; Garg, M. Dengue, Zika and chikungunya: Emerging arboviruses in the new world. *West. J. Emerg. Med.* **2016**, *17*, 671–679. [CrossRef] [PubMed]
138. Mordecai, E.A.; Cohen, J.M.; Evans, M.V.; Gudapati, P.; Johnson, L.R.; Lippi, C.A.; Miazgowicz, K.; Murdock, C.C.; Rohr, J.R.; Ryan, S.J.; et al. Detecting the impact of temperature on transmission of Zika, dengue, and chikungunya using mechanistic models. *PLoS Negl. Trop. Dis.* **2017**, *11*, e0005568. [CrossRef] [PubMed]
139. Tjaden, N.B.; Thomas, S.M.; Fischer, D.; Beierkuhnlein, C. Extrinsic incubation period of dengue: Knowledge, backlog, and applications of temperature dependence. *PLoS Negl. Trop. Dis.* **2013**, *7*, e2207. [CrossRef]
140. Rückert, C.; Weger-Lucarelli, J.; Garcia-Luna, S.M.; Young, M.C.; Byas, A.D.; Murrieta, R.A.; Fauver, J.R.; Ebel, G.D. Impact of simultaneous exposure to arboviruses on infection and transmission by Aedes aegypti mosquitoes. *Nat. Commun.* **2017**, *8*, 15412. [CrossRef] [PubMed]
141. Göertz, G.P.; Vogels, C.B.F.; Geertsema, C.; Koenraadt, C.J.M.; Pijlman, G.P. Mosquito co-infection with Zika and chikungunya virus allows simultaneous transmission without affecting vector competence of Aedes aegypti. *PLoS Negl. Trop. Dis.* **2017**, *11*, e0005654. [CrossRef] [PubMed]
142. Dejnirattisai, W.; Supasa, P.; Wongwiwat, W.; Rouvinski, A.; Barba-Spaeth, G.; Duangchinda, T.; Sakuntabhai, A.; Cao-Lormeau, V.M.; Malasit, P.; Rey, F.A.; et al. Dengue virus sero-cross-reactivity drives antibody-dependent enhancement of infection with Zika virus. *Nat. Immunol.* **2016**, *17*, 1102–1108. [CrossRef] [PubMed]

143. Screaton, G.; Mongkolsapaya, J.; Yacoub, S.; Roberts, C. New insights into the immunopathology and control of dengue virus infection. *Nat. Rev. Immunol.* **2015**, *15*, 745–759. [CrossRef]
144. Weiskopf, D.; Angelo, M.A.; de Azeredo, E.L.; Sidney, J.; Greenbaum, J.A.; Fernando, A.N.; Broadwater, A.; Kolla, R.V.; De Silva, A.D.; de Silva, A.M.; et al. Comprehensive analysis of dengue virus-specific responses supports an HLA-linked protective role for CD8$^+$ T cells. *Proc. Natl. Acad. Sci. USA* **2013**, *110*, E2046–E2053. [CrossRef] [PubMed]
145. Rivino, L. T cell immunity to dengue virus and implications for vaccine design. *Expert Rev. Vaccines* **2016**, *15*, 443–453. [CrossRef]
146. Grifoni, A.; Pham, J.; Sidney, J.; O'Rourke, P.H.; Paul, S.; Peters, B.; Martini, S.R.; de Silva, A.D.; Ricciardi, M.J.; Magnani, D.M.; et al. Prior dengue virus exposure shapes T cell immunity to Zika virus in humans. *J. Virol.* **2017**, *91*, e01469-17. [CrossRef] [PubMed]
147. Elong Ngono, A.; Shresta, S. Immune response to dengue and Zika. *Annu. Rev. Immunol.* **2018**, *36*, 279–308. [CrossRef] [PubMed]
148. Weiskopf, D.; Cerpas, C.; Angelo, M.A.; Bangs, D.J.; Sidney, J.; Paul, S.; Peters, B.; Sanches, F.P.; Silvera, C.G.; Romero-Sanchez, C.; et al. Human CD8+ T-cell responses against the 4 dengue virus serotypes are associated with distinct patterns of protein targets. *J. Infect. Dis.* **2015**, *212*, 1743–1751. [CrossRef] [PubMed]
149. Yauch, L.E.; Shresta, S. Mouse models of dengue virus infection and disease. *Antivir. Res.* **2008**, *80*, 87–93. [CrossRef] [PubMed]
150. Wen, J.; Elong Ngono, A.; Regla-Nava, J.A.; Kim, K.; Gorman, M.J.; Diamond, M.S.; Shresta, S. Dengue virus-reactive CD8+ T cells mediate cross-protection against subsequent Zika virus challenge. *Nat. Commun.* **2017**, *8*, 1459. [CrossRef] [PubMed]
151. Ricciardi, M.J.; Magnani, D.M.; Grifoni, A.; Kwon, Y.C.; Gutman, M.J.; Grubaugh, N.D.; Gangavarapu, K.; Sharkey, M.; Silveira, C.G.T.; Bailey, V.K.; et al. Ontogeny of the B- and T-cell response in a primary Zika virus infection of a dengue-naïve individual during the 2016 outbreak in Miami, FL. *PLoS Negl. Trop. Dis.* **2017**, *11*, e0006000. [CrossRef] [PubMed]
152. Wauquier, N.; Becquart, P.; Nkoghe, D.; Padilla, C.; Ndjoyi-Mbiguino, A.; Leroy, E.M. The acute phase of chikungunya virus infection in humans is associated with strong innate immunity and T CD8 cell activation. *J. Infect. Dis.* **2011**, *204*, 115–123. [CrossRef] [PubMed]
153. Lum, F.M.; Teo, T.H.; Lee, W.W.L.; Kam, Y.W.; Rénia, L.; Ng, L.F.P. An essential role of antibodies in the control of chikungunya virus infection. *J. Immunol.* **2013**, *190*, 6295–6302. [CrossRef] [PubMed]
154. Keasler, C.; Rodriguez, A.; Rodriguez, M.; Quinonez, E.; Barrera, R. Increased efficiency of Aedes aegypti (Diptera: Culicidae) surveillance through mobile technology and improved deployment of BGS traps. *J. Med. Entomol.* **2019**, *56*, 1395–1405. [CrossRef]
155. Bowman, L.R.; Donegan, S.; McCall, P.J. Is dengue vector control deficient in effectiveness or evidence?: Systematic review and meta-analysis. *PLoS Negl. Trop. Dis.* **2016**, *10*, e0004551. [CrossRef] [PubMed]
156. Wilson, A.L.; Courtenay, O.; Kelly-Hope, L.A.; Scott, T.W.; Takken, W.; Torr, S.J.; Lindsay, S.W. The importance of vector control for the control and elimination of vector-borne diseases. *PLoS Negl. Trop. Dis.* **2020**, *14*, e0007831. [CrossRef] [PubMed]
157. Utarini, A.; Indriani, C.; Ahmad, R.A.; Tantowijoyo, W.; Arguni, E.; Ansari, M.R.; Supriyati, E.; Wardana, D.S.; Meitika, Y.; Ernesia, I.; et al. Efficacy of Wolbachia-infected mosquito deployments for the control of dengue. *N. Engl. J. Med.* **2021**, *384*, 2177–2186. [CrossRef] [PubMed]
158. Abbas, K.; Procter, S.R.; van Zandvoort, K.; Clark, A.; Funk, S.; Mengistu, T.; Hogan, D.; Dansereau, E.; Jit, M.; Flasche, S.; et al. Routine childhood immunisation during the COVID-19 pandemic in Africa: A benefit-risk analysis of health benefits versus excess risk of SARS-CoV-2 infection. *Lancet Glob. Health* **2020**, *8*, e1264–e1272. [CrossRef]
159. Kruk, M.E.; Myers, M.; Varpilah, S.T.; Dahn, B.T. What is a resilient health system? Lessons from Ebola. *Lancet* **2015**, *385*, 1910–1912. [CrossRef]
160. Hanefeld, J.; Mayhew, S.; Legido-Quigley, H.; Martineau, F.; Karanikolos, M.; Blanchet, K.; Liverani, M.; Yei Mokuwa, E.; McKay, G.; Balabanova, D. Towards an understanding of resilience: Responding to health systems shocks. *Health Policy Plan.* **2018**, *33*, 355–367. [CrossRef] [PubMed]
161. Henning, K.J. What is syndromic surveillance? *MMWR Suppl.* **2004**, *53*, 5–11. [PubMed]
162. Launiala, A. How much can a KAP survey tell us about people's knowledge, attitudes and practices? Some observations from medical anthropology research on malaria in pregnancy in Malawi. *Anthropol. Matters* **2009**, *11*, 1–13. [CrossRef]
163. Parks, W.; Lloyd, L. *Planning Social Mobilization and Communication for Dengue Fever Prevention and Control: A Step-by-Step Guide*; World Health Organization: Geneva, Switzerland, 2004.
164. Roberton, T.; Carter, E.D.; Chou, V.B.; Stegmuller, A.R.; Jackson, B.D.; Tam, Y.; Sawadogo-Lewis, T.; Walker, N. Early estimates of the indirect effects of the COVID-19 pandemic on maternal and child mortality in low-income and middle-income countries: A modelling study. *Lancet Glob. Health* **2020**, *8*, e901–e908. [CrossRef]
165. Christophers, S.R. *Aedes aegypti (L.) the Yellow Fever Mosquito: Its Life History, Bionomics and Structure*; Cambridge University Press: Cambridge, UK, 1960.
166. Scott, T.W.; Amerasinghe, P.H.; Morrison, A.C.; Lorenz, L.H.; Clark, G.G.; Strickman, D.; Kittayapong, P.; Edman, J.D. Longitudinal studies of Aedes aegypti (Diptera: Culicidae) in Thailand and Puerto Rico: Blood feeding frequency. *J. Med. Entomol.* **2000**, *37*, 89–101. [CrossRef] [PubMed]

167. Reiter, P. Oviposition, dispersal, and survival in Aedes aegypti: Implications for the efficacy of control strategies. *Vector Borne Zoonotic Dis.* **2007**, *7*, 261–273. [CrossRef]
168. Colón-González, F.J.; Sewe, M.O.; Tompkins, A.M.; Sjödin, H.; Casallas, A.; Rocklöv, J.; Caminade, C.; Lowe, R. Projecting the risk of mosquito-borne diseases in a warmer and more populated world: A multi-model, multi-scenario intercomparison modelling study. *Lancet Planet. Health* **2021**, *5*, e404–e414. [CrossRef] [PubMed]
169. Peeling, R.W.; Boeras, D.I.; Nkengasong, J. Re-imagining the future of diagnosis of neglected tropical diseases. *Comput. Struct. Biotechnol. J.* **2017**, *15*, 271–274. [CrossRef] [PubMed]
170. Boehme, C.C.; Nabeta, P.; Hillemann, D.; Nicol, M.P.; Shenai, S.; Krapp, F.; Allen, J.; Tahirli, R.; Blakemore, R.; Rustomjee, R.; et al. Rapid molecular detection of tuberculosis and rifampin resistance. *N. Engl. J. Med.* **2010**, *363*, 1005–1015. [CrossRef] [PubMed]
171. Pybus, O.G.; Suchard, M.A.; Lemey, P.; Bernardin, F.J.; Rambaut, A.; Crawford, F.W.; Gray, R.R.; Arinaminpathy, N.; Stramer, S.L.; Busch, M.P.; et al. Unifying the spatial epidemiology and molecular evolution of emerging epidemics. *Proc. Natl. Acad. Sci. USA* **2012**, *109*, 15066–15071. [CrossRef] [PubMed]
172. Shepard, D.S.; Undurraga, E.A.; Halasa, Y.A.; Stanaway, J.D. The global economic burden of dengue: A systematic analysis. *Lancet Infect. Dis.* **2016**, *16*, 935–941. [CrossRef] [PubMed]
173. Morse, S.S.; Mazet, J.A.K.; Woolhouse, M.; Parrish, C.R.; Carroll, D.; Karesh, W.B.; Zambrana-Torrelio, C.; Lipkin, W.I.; Daszak, P. Prediction and prevention of the next pandemic zoonosis. *Lancet* **2012**, *380*, 1956–1965. [CrossRef] [PubMed]
174. Jones, B.A.; Grace, D.; Kock, R.; Alonso, S.; Rushton, J.; Said, M.Y.; McKeever, D.; Mutua, F.; Young, J.; McDermott, J.; et al. Zoonosis emergence linked to agricultural intensification and environmental change. *Proc. Natl. Acad. Sci. USA* **2013**, *110*, 8399–8404. [CrossRef] [PubMed]
175. Bambrick, H.J.; Woodruff, R.E.; Hanigan, I.C. Climate change could threaten blood supply by altering the distribution of vector-borne disease: An Australian case-study. *Glob. Health Action* **2009**, *2*, 2059. [CrossRef] [PubMed]
176. World Health Organization. *Global Vector Control Response 2017–2030*; WHO Press: Geneva, Switzerland, 2017.
177. Holmes, E.C.; Twiddy, S.S. The origin, emergence and evolutionary genetics of dengue virus. *Infect. Genet. Evol.* **2003**, *3*, 19–28. [CrossRef] [PubMed]

viruses

MDPI

Article

Unexpected Predictors of Mortality During a DENV-3 Outbreak in Western Mexico: Seizures, Polyserositis, and Renal Dysfunction Without Severe Thrombocytopenia

Martha A. Mendoza-Hernandez [1,2], Janet Diaz-Martinez [3], Gustavo A. Hernández-Fuentes [2,4,5], Fabian Rojas-Larios [2], Katya A. Cárdenas-Cárdenas [1], Paulina García de León-Flores [1], David A. Rojas-Cruz [1], Roberto Aceves-Calvario [1], Ernesto Gómez-Sandoval [1], Montserrat Árciga-García [1], José Guzmán-Esquivel [6], Valery Melnikov [2], Francisco Espinoza-Gómez [7] and Iván Delgado-Enciso [2,5,8,*]

1. Hospital General de Manzanillo IMSS-Bienestar, Av. Elías Zamora Verduzco S/N, Nuevo Salagua, Manzanillo City 28869, Mexico; mendoza_martha@ucol.mx (M.A.M.-H.); katyacardenas12@gmail.com (K.A.C.-C.); pau.g96@hotmail.com (P.G.d.L.-F.); drojas3@ucol.mx (D.A.R.-C.); roberto_aceves0@ucol.mx (R.A.-C.); gosaer480@icloud.com (E.G.-S.); rabe_m21@hotmail.com (M.Á.-G.)
2. Department of Molecular Medicine, School of Medicine, University of Colima, Colima City 28040, Mexico; gahfuentes@gmail.com (G.A.H.-F.); frojas@ucol.mx (F.R.-L.); valery.melnikoff@gmail.com (V.M.)
3. Research Center in Minority Institutions, Department of Dietetics and Nutrition, Florida International University (FIU-RCMI), Miami, FL 33199, USA; jdimarti@fiu.edu
4. Faculty of Chemical Sciences, University of Colima, Coquimatlan City 28400, Mexico
5. State Cancerology Institute of Colima, Health Services of the Mexican Social Security Institute for Welfare (IMSS-BIENESTAR), Colima City 28085, Mexico
6. Clinical Epidemiology Research Unit, Mexican Institute of Social Security, Villa de Alvarez 28984, Mexico; jose.esquivel@imss.gob.mx
7. Regional University Hospital (IMSS-Bienestar), Health Services of the State of Colima, Colima City 28060, Mexico; fespin@ucol.mx
8. Robert Stempel College of Public Health and Social Work, Florida International University, Miami, FL 33199, USA

* Correspondence: ivan_delgado_enciso@ucol.mx; Tel.: +52-312-3161099

Abstract: Dengue mortality has traditionally been associated with severe thrombocytopenia and hemorrhagic complications. However, during 2024, dengue virus serotype 3 (DENV-3) increased significantly in western Mexico, leading to the emergence of a distinct clinical pattern. We conducted a retrospective cohort study of hospitalized dengue patients at the General Hospital of Colima (January–August 2024). Clinical features, laboratory parameters, and outcomes were compared between survivors and non-survivors. Among 201 hospitalized patients, 6 (3.0%) died. All deceased patients presented with generalized seizures, polyserositis (pleural effusion and/or ascites), and required mechanical ventilation. Contrary to classical patterns, they did not have severe thrombocytopenia. Instead, they showed significantly higher white blood cell counts and notably increased levels of serum urea and BUN, suggesting early renal impairment. ROC analysis indicated that BUN (AUC 0.904) and urea (AUC 0.906) were good to excellent discriminators of mortality. During 2024, with an increase in DENV-3 circulation, mortality was associated with neurological and systemic complications, including seizures and polyserositis, as well as biochemical evidence of renal dysfunction—but not with severe thrombocytopenia. These findings challenge current paradigms and highlight the need for early recognition of atypical clinical patterns.

Keywords: dengue virus serotype 3 (DENV-3); neurological complications; renal dysfunction; polyserositis; epidemiology

1. Introduction

Dengue virus infection is a major and growing public health concern worldwide, with an estimated 390 million infections and 25,000 deaths annually, predominantly in tropical and subtropical regions [1,2]. In the Americas, the burden of dengue reached record levels in 2024, with Mexico experiencing one of the most severe epidemics in its history [3]. Notably, the state of Colima has emerged as one of the most affected regions in Mexico, reporting the highest national incidence rate—over 622 cases per 100,000 inhabitants in 2024—amid a dramatic surge in cases [4]. This epidemiological scenario has placed unprecedented strain on local healthcare systems and highlighted the urgent need for improved clinical management and public health strategies.

Historically, severe dengue has been linked to plasma leakage and hemorrhagic complications, primarily driven by severe thrombocytopenia [5]. Consistently, the WHO classification highlights warning signs such as persistent vomiting, mucosal bleeding, and a rapid drop in platelet count as predictors of severe disease [6,7]. However, recent outbreaks have challenged these classical paradigms. Emerging evidence suggests that the clinical spectrum of severe dengue is broader and more heterogeneous than previously recognized, with increasing reports of atypical presentations—including neurological involvement, acute organ dysfunction, and systemic inflammatory responses—particularly in adults and patients with comorbidities [8].

The 2024 epidemic in Mexico has been characterized by a predominance of dengue virus serotype 3 (DENV-3) [9], a strain historically associated with more severe disease manifestations and increased risk of outbreaks in previously exposed populations [10]. Furthermore, the interplay between host factors (such as comorbidities and immune status), viral characteristics, and environmental determinants is increasingly recognized as central to understanding the pathogenesis and outcomes of dengue infection [11]. Recent studies have highlighted the potential prognostic value of laboratory markers beyond thrombocytopenia, including leukocyte indices, markers of renal function, and inflammatory ratios, in identifying patients at highest risk of adverse outcomes [12–14].

Against this backdrop, there is a pressing need to refine risk stratification tools and clinical management algorithms for dengue, particularly in high-incidence settings. A deeper understanding of the constantly evolving clinical and biochemical predictors of mortality is essential to optimize patient outcomes, efficiently allocate resources, and inform public health policies during epidemic emergencies. Therefore, this paper reports the clinical outcomes of hospitalized patients in Colima, Mexico, emphasizing the signs and parameters of those who died.

2. Materials and Methods

2.1. Study Design

A retrospective observational study was carried out using data from a cohort of patients hospitalized with confirmed DENV-3 infection during the 2024 epidemic in a tropical/coastal city in the state of Colima, Mexico. The study was conducted at the General Hospital of Manzanillo IMSS-Bienestar "Hospital General de Manzanillo IMSS-Bienestar" as part of a project that analyzes the epidemiology of dengue over time. It was approved by the Central HRU (Regional University Hospital of Colima "Hospital Regional Universitario de Colima") Research Ethics Committee (approval code: CI 2024/2/SR/MI/254). This study adhered to the Strengthening the Reporting of Observational Studies in Epidemiology (STROBE) guidelines. As it was a retrospective analysis based on medical records, the ethics committee waived the requirement for informed consent [15]. Nonetheless, strict measures were taken to ensure the anonymity and confidentiality of all patient data.

Inclusion criteria comprised (1) age $\geq$ 16 years; (2) laboratory-confirmed dengue infection, defined as positive NS1 antigen and/or IgM/IgG serology and/or PCR, according to national guidelines; and (3) hospitalization for clinical management. Patients with incomplete clinical records or who were discharged against medical advice were excluded.

2.2. Data Collection and Outcomes

Data were extracted manually from electronic medical records and archived laboratory databases by two independent investigators using a structured case report form. The primary outcome was in-hospital mortality, defined as death occurring during hospitalization due to any cause. Patients were stratified post hoc into survivors and deceased for comparative analyses. Others variables were collected for analysis: demographic and comorbidity data, including age, sex, and the presence of diabetes mellitus, hypertension, obesity, smoking status, autoimmune diseases, HIV infection, ischemic heart or cerebrovascular disease, and chronic kidney disease; pharmacological treatments administered during hospitalization, including the use of paracetamol, antihypertensive agents, insulin, corticosteroids, non-steroidal anti-inflammatory drugs (NSAIDs), and antibiotics; and laboratory parameters obtained during hospitalization.

Hematological parameters included hemoglobin, hematocrit, red cell distribution width (RDW), leukocyte count, and differential counts (neutrophils, lymphocytes, monocytes, eosinophils, and basophils). Absolute white blood cell counts (neutrophils, lymphocytes, monocytes, eosinophils, and basophils) were reported in number of cells per microliter ($\times 10^3/\mu L$), as determined by an automated hematology analyzer (DxH 690T, Beckman Coulter, Brea, CA, USA). These values represent the actual cell count, in contrast to relative percentages, and allow for a more accurate assessment in the context of leukopenia, which is common in dengue infection [16,17].

Biochemical analyses were performed using a fully automated dry chemistry analyzer (VITROS 5600 Integrated System, Ortho Clinical Diagnostics, Rochester, NY, USA), and included serum glucose, blood urea nitrogen (BUN), urea, creatinine, total, direct, and indirect bilirubin, and liver enzymes including aspartate aminotransferase (AST), alanine aminotransferase (ALT), gamma-glutamyl transferase (GGT), alkaline phosphatase (ALP), lactate dehydrogenase (LDH), and serum albumin. Liver function tests and biochemical panels were also processed using the VITROS 5600 platform. Coagulation parameters included prothrombin time (PT), activated partial thromboplastin time (aPTT), and international normalized ratio (INR), and were assessed using an automated coagulometer (ACL TOP 550 CTS, Instrumentation Laboratory, Bedford, MA, USA).

All laboratory analyses were performed at the clinical laboratory of Hospital of Manzanillo IMSS-Bienestar, located in Colima, Mexico, and were conducted under standardized protocols and internal quality controls.

2.3. Statistical Analysis

Continuous variables were tested for normality using the Shapiro–Wilk test. Since most variables displayed non-normal distribution, they are reported as medians with 25th and 75th percentiles (Q1 and Q2). Categorical variables are presented as frequencies and percentages. Comparisons between survivors and non-survivors were made using the Mann–Whitney U test for continuous variables and with Fisher's exact test for categorical variables. Receiver operating characteristic (ROC) curve analysis was used to evaluate the discriminative ability of selected clinical and biochemical parameters in predicting in-hospital mortality among patients with dengue. All evaluated variables were included in this analysis to comprehensively explore their potential prognostic value. Although not all parameters reached statistical significance, they are retained in the results to provide a complete profile and to facilitate future research [18,19].

Given the small sample size, we opted not to report sensitivity, specificity, or optimal cutoff points, as these values may lack robustness and clinical reliability and lead to overfitting and misleading conclusions. Instead, we focused on reporting the area under the curve (AUC) with corresponding 95% confidence intervals (CIs) to explore the potential utility of these markers as prognostic indicators. AUC values were interpreted as follows: >0.90 = excellent, 0.80–0.89 = good, between 0.70 and 0.79 as fair, and between 0.50 and 0.69 as poor [20]. The aim was to preliminarily assess whether these variables show potential discriminatory value, as reflected by the AUC [21]. A p-value < 0.05 was considered statistically significant. Analyses were performed using SPSS Statistics version 20 software program (IBM Corp., Armonk, NY, USA) [22].

2.4. Quality Control

Double data entry and verification were employed to minimize transcription errors. Inconsistencies were resolved through review by a third investigator. Missing data were handled by pairwise deletion and not imputed, due to the retrospective design and low frequency of missingness.

3. Results

A total of 196 patients hospitalized with laboratory-confirmed dengue infection were included in the analysis, of whom 6 (3.1%) died during hospitalization. The minimum and maximum ages were 16 and 87 years, respectively. Table 1 summarizes the main demographic and clinical characteristics of the patients, stratified by vital status at hospital discharge. Patients who died were significantly older than those who survived (median age 53 [IQR 28–65] vs. 27 years [18–39]; $p = 0.036$). Most patients who died were men, although a significant difference was not reached (83.3% vs. 49.7, $p = 0.212$). Comorbidities such as diabetes mellitus (50.0% vs. 10.0%; $p = 0.020$) and hypertension (50.0% vs. 7.2%; $p = 0.008$) were significantly more prevalent among those who died. Obesity, smoking, autoimmune diseases, and chronic kidney disease showed no significant differences between groups. Regarding pharmacological management, paracetamol was administered to 100% of deceased patients and 65.6% of survivors ($p = 0.182$). Notably, antihypertensive drugs were more frequently used in patients who died (33.3% vs. 5.2%; $p = 0.045$), consistent with their higher burden of cardiovascular comorbidities.

Table 1. Baseline characteristics and medications administered to hospitalized patients with dengue, stratified by vital status at discharge.

Variable	Survivors (n = 190)	Deceased (n = 6)	p-Value
Age (years)	27 (20–41)	53 (28–65)	0.036
Female	50.3%	16.7%	0.212
Obesity	7.2%	0.0%	0.645
Diabetes mellitus (DM)	10.0%	50.0%	0.020
Hypertension (HTN)	7.2%	50.0%	0.008
Smoking	11.0%	16.7%	0.513
Pulmonary disease	1.0%	0.0%	0.945
Autoimmune disease	0.5%	0.0%	0.972
HIV	1.4%	0.0%	0.921
Ischemic disease	0.5%	0.0%	0.972
Chronic kidney disease	1.0%	0.0%	0.945
IgG positive	82.5%	80.0%	0.999

Table 1. *Cont.*

Variable	Survivors (n = 190)	Deceased (n = 6)	p-Value
Medications Administered			
Paracetamol	65.6%	100.0%	0.182
Antibiotics	1.4%	0.0%	0.918
NSAIDs	0.0%	0.0%	
Steroids	1.0%	0.0%	0.999
Antihypertensives	5.2%	33.3%	0.045
Insulin	1.6%	0.0%	0.911

A comparison between groups was performed using Fisher's exact test for categorical variables. Age is presented as the median (25th and 75th percentiles) and was compared using a one-tailed Mann–Whitney U test, under the hypothesis that "those who died were older than those who survived".

3.1. Laboratory Findings Prior to Discharge

Table 2 compares hematological and biochemical parameters between survivors and deceased patients, focusing on values obtained during the final two days of hospitalization. Deceased patients had significantly lower hemoglobin levels (median 9.95 g/dL vs. 13.80 g/dL [12.70–15.10]; $p < 0.001$) and hematocrit (27.95% [23.63–32.63] vs. 39.30% [36.50–42.30]; $p < 0.001$). Red cell distribution width (RDW) was significantly higher among deceased patients ($p = 0.007$), suggesting greater erythrocyte anisocytosis.

Table 2. Comparison of hematological and biochemical parameters between survivors and deceased patients with dengue during final two days of hospitalization.

Parameter	Survivors			Non-Survivors			p-Value
	Median	Q1	Q3	Median	Q1	Q3	
Hemoglobin (g/dL)	13.80	12.70	15.10	9.95	8.28	11.75	<0.001
Hematocrit (%)	39.30	36.50	42.30	27.95	23.63	32.63	<0.001
RDW-CV (%)	12.70	12.10	13.50	13.20	12.70	14.00	0.007
Platelets (×103/μL)	50.00	35.00	66.00	94.50	48.75	115.50	<0.001
Mean Platelet Volume (MPV, fL)	11.85	11.30	12.60	12.70	11.08	13.20	0.575
White Blood Cells (×103/μL)	5.07	3.81	6.94	9.77	4.55	13.29	<0.001
Absolute Neutrophils (×103/μL)	1.57	1.00	2.43	6.99	2.61	11.17	<0.001
Absolute Lymphocytes (×103/μL)	2.60	1.90	3.49	1.25	1.11	1.74	<0.001
Absolute Monocytes (×103/μL)	0.46	0.34	0.63	0.54	0.32	0.98	<0.001
Absolute Eosinophils (×103/μL)	0.05	0.02	0.11	0.07	0.01	0.15	0.227
Absolute Basophils (×103/μL)	0.04	0.03	0.06	0.03	0.01	0.04	0.122
Glucose (mg/dL)	91.00	84.50	106.00	96.00	84.00	123.00	0.004
BUN (mg/dL)	7.00	6.00	10.00	16.00	12.75	62.50	<0.001
Urea (mg/dL)	14.98	12.84	21.40	34.24	27.29	133.71	<0.001
Creatinine (mg/dL)	0.70	0.60	0.80	0.80	0.60	1.50	<0.001
Direct Bilirubin (mg/dL)	0.38	0.29	0.50	0.40	0.21	2.96	0.163
Indirect Bilirubin (mg/dL)	0.29	0.18	0.41	0.29	0.13	0.38	0.922
Total Bilirubin (mg/dL)	0.68	0.50	0.86	0.69	0.34	3.32	0.164
AST (U/L)	234.00	135.50	387.50	125.50	74.75	271.75	0.001
ALT (U/L)	156.00	90.75	250.00	75.50	31.00	219.75	<0.001
GGT (U/L)	209.00	98.50	359.00	102.50	42.00	308.75	0.078
Alkaline Phosphatase (U/L)	95.00	65.00	141.00	149.50	38.25	243.50	0.638
LDH (U/L)	529.50	404.75	848.50	551.00	322.00	1180.50	0.072
Albumin (g/dL)	3.30	3.00	3.60	2.00	1.85	2.05	<0.001
Prothrombin Time (PT, s)	11.00	10.50	11.70	12.50	11.30	12.43	0.011
aPTT (s)	36.70	33.00	42.00	30.60	28.00	32.75	<0.001
INR	0.95	0.90	1.01	1.08	0.97	1.19	0.010

Data are shown as the median (Q1–Q3). All comparisons made using the non-parametric Mann–Whitney U test. BUN: blood urea nitrogen; AST: aspartate aminotransferase; ALT: alanine aminotransferase; GGT: gamma-glutamyl transferase; PT: prothrombin time; aPTT: activated partial thromboplastin time; INR: international normalized ratio.

Unexpectedly, platelet counts were significantly higher in the deceased group (median $94.5 \times 10^3/\mu L$ [48.75–115.50] vs. $50.0 \times 10^3/\mu L$ [35.00–66.00]; $p < 0.001$).

Markers of renal function, including blood urea nitrogen (BUN), urea, and creatinine, were significantly worse in deceased patients (all $p < 0.001$), suggesting that acute kidney injury may have contributed to mortality. Serum albumin levels were also significantly lower in this group (median 2.00 g/dL vs. 3.30 g/dL; $p < 0.001$), consistent with hypoalbuminemia, which is associated with systemic inflammation and capillary leakage.

Although transaminase levels (AST and ALT) were elevated in both groups, they were paradoxically lower in deceased patients ($p = 0.001$ and $p < 0.001$, respectively), possibly reflecting hepatic exhaustion or variability in sampling timing during the terminal phase of illness.

Coagulation parameters, including prothrombin time (PT), international normalized ratio (INR), and activated partial thromboplastin time (aPTT), showed statistically significant differences between groups; however, these changes did not appear to be clinically relevant.

3.2. Blood Markers with Potential Predictive Value for Mortality

A receiver operating characteristic (ROC) curve analysis was conducted to evaluate the ability of various clinical and biochemical parameters to discriminate mortality among hospitalized patients with dengue. Several markers showed good to excellent discriminatory capacity, particularly urea (AUC: 0.906; 95% CI: 0.808–1.000; $p < 0.001$), blood urea, and nitrogen (BUN; AUC: 0.904; 95% CI: 0.805–1.000; $p < 0.001$). Other parameters, such as platelet count and absolute neutrophil count, demonstrated moderate predictive performance. The full results are summarized in Table 3.

Table 3. Discriminatory capacity (AUC) of clinical and laboratory parameters for mortality prediction in patients hospitalized with dengue.

Parameter	AUC	CI 95%		*p*-Value
		Lower	Upper	
Hemoglobin (HG)	0.141	0.000	0.303	<0.001
Hematocrit (HTO)	0.129	0.000	0.290	<0.001
RDW-CV	0.646	0.519	0.772	0.100
Platelets	0.799	0.623	0.976	0.001
Mean Platelet Volume (MPV)	0.638	0.421	0.855	0.136
Leukocytes	0.757	0.576	0.939	0.006
Absolute Neutrophils	0.835	0.671	0.998	<0.001
Absolute Lymphocytes	0.166	0.060	0.271	<0.001
Absolute Monocytes	0.605	0.379	0.831	0.259
Absolute Eosinophils	0.571	0.374	0.769	0.442
Absolute Basophils	0.353	0.159	0.547	0.113
Glucose	0.519	0.313	0.726	0.835
Blood Urea Nitrogen (BUN)	0.904	0.805	1.000	<0.001
Urea	0.906	0.808	1.000	<0.001
Creatinine	0.624	0.427	0.822	0.180
Direct Bilirubin (DB)	0.547	0.249	0.846	0.691
Indirect Bilirubin (IB)	0.448	0.236	0.661	0.664
Total Bilirubin (TB)	0.528	0.217	0.839	0.815
AST (TGO)	0.266	0.105	0.428	0.002
ALT (TGP)	0.237	0.085	0.389	0.001
GGT	0.418	0.298	0.538	0.279
Alkaline Phosphatase (ALP)	0.571	0.383	0.760	0.346
LDH	0.359	0.228	0.490	0.063
Albumin	0.106	0.000	0.249	<0.001
Prothrombin Time (PT)	0.767	0.663	0.870	0.010
aPTT	0.066	0.014	0.118	<0.001
INR	0.767	0.648	0.882	0.010

The area under the curve (AUC), 95% confidence intervals (CIs), and *p*-values are presented for each variable. Due to the limited sample size, optimal cutoff points, sensitivity, and specificity values were not included to avoid potential overestimation.

3.3. Key Clinical Differences Between Survivors and Non-Survivors

Notable differences in clinical presentation were observed between patients who survived and those who died during hospitalization for dengue. All non-survivors (6/6; 100%) required invasive mechanical ventilation (IMV) and experienced seizures, while none of the survivors (0/190; 0%) had these complications ($p < 0.001$ for both). Likewise, polyserositis—assessed in a subset of patients who underwent thoracoabdominal ultrasound ($n = 135$)—was present only among those who died (6/6; 100%) and absent in survivors (0/129; 0%) ($p < 0.001$). No significant difference was found in the presence of hepatomegaly between the two groups (1/129 survivors, 0.8%; 0/6 non-survivors, 0%; $p = 0.956$). These findings outline clinical severity in fatal cases. Comparisons were performed using Fisher's exact test.

4. Discussion

Currently, Mexico, as well as several Latin American countries are facing a significant dengue outbreak, with some regions reporting record-breaking numbers of cases. In the last decade, DENV-2 predominated in Mexico, but in 2023 and especially 2024, the prevalence of DENV-3 cases increased considerably, leading to the emergence of atypical clinical presentations, with greater liver involvement, as well as respiratory symptoms that simulate COVID-19 or other respiratory viruses, with normal or slightly decreased platelet counts [23]. However, this is an underreported phenomenon. During 2024, the Colima region had the highest incidence of confirmed dengue cases in Mexico (622 confirmed cases per 100,000 inhabitants) [4]. The 2024 outbreak in Colima, Mexico, has unveiled an atypical clinical and biochemical profile associated with in-hospital mortality, prompting critical reconsideration of existing paradigms for dengue severity assessment. Typically, severe thrombocytopenia and hemorrhagic manifestations have been considered hallmark predictors of poor outcomes in dengue fever, particularly in previous studies [24]. However, the findings of the present study demonstrate that during this outbreak, mortality was not linked to classical hemorrhagic complications or profound thrombocytopenia. Instead, neurological events (notably generalized seizures), systemic inflammatory features (such as polyserositis), and early renal dysfunction emerged as the most salient predictors of mortality.

Among the six patients who died, they were mostly men, were significantly older than the survivors (median age 53 vs. 27 years), and exhibited a higher prevalence of diabetes mellitus and hypertension. The gender difference in mortality, although not statistically significant, is consistent with previous reports showing higher mortality in men [25]. Nonetheless, prior studies have highlighted that age and comorbidities modulate the host immune response, potentially exacerbating dengue pathophysiology. A multicenter study in Mexico [26] also reported that patients with diabetes and hypertension were more likely to develop severe dengue and succumb to complications [27], possibly due to chronic endothelial dysfunction and impaired immune regulation. Similar associations were observed during DENV-2 and DENV-1 outbreaks in other regions of America [28–30]. Diabetes mellitus has been implicated in amplifying dengue-associated plasma leakage and cytokine storms [31–33]. Hyperglycemia is known to upregulate pro-inflammatory mediators and compromise leukocyte function, leading to an exaggerated yet ineffective immune response [34]. Hypertension, on the other hand, may predispose microvascular damage and exacerbate capillary leakage, culminating in multiorgan dysfunction [35].

Contrary to long-standing beliefs that low platelet counts are central to dengue mortality, our data demonstrate that non-survivors exhibited significantly higher platelet counts (median $94.5 \times 10^3/\mu L$) than survivors (median $50.0 \times 10^3/\mu L$). This unexpected s inversion may reflect distinct immunopathological dynamics associated with DENV-3 circulation in this region. Historically, outbreaks caused by DENV-2 and DENV-1 have shown stronger associations with platelet consumption, disseminated intravascular coagulation (DIC), and bleeding [36,37]. In contrast, DENV-3, particularly certain genotypes, has demonstrated a higher neurotropic potential in animal models [38], possibly indicating a different pathophysiological profile. However, this finding should be interpreted with caution due to the limited sample size of fatal cases and the single-center design of our study. It is possible that regional factors, timing of sample collection, or viral genotypic variations influenced the observed results; for that reason, more studies are needed with broader populations and different epidemic contexts.

This neurotropism has also been reported in patients, primarily with DENV-2 and DENV-3 infections [39], although cases involving DENV-1 have also been documented [33]. The relatively preserved platelet counts observed in non-survivors may indicate a distinct pathophysiological pathway in which mortality is driven not by hemorrhagic complications but rather by inflammatory and organ-specific mechanisms—particularly those affecting the brain, lungs, and kidneys. Notably, none of the deceased exhibited hemorrhagic symptoms.

All patients who died developed generalized seizures, a feature not typically emphasized in standard dengue severity classifications [40]. Although neurological manifestations of dengue are increasingly recognized—including encephalopathy, encephalitis, and seizures—they remain underreported in most clinical settings [41]. A study from India has proposed that neuro-dengue can occur due to (1) direct neurotropism (encephalitis, meningitis, myelitis, and myositis); (2) systemic complications (encephalopathy, stroke, and hypokalemic paralysis); (3) post-infectious/immune-mediated mechanisms (acute disseminated encephalomyelitis (ADEM), Guillain–Barré syndrome, and optic neuritis) [42,43]. The universal presence of seizures in all fatal cases in the present report suggests a profound neurotropic effect or secondary neuroinflammation, possibly exacerbated by renal dysfunction and uremic encephalopathy. The elevated BUN and urea levels in these patients support this hypothesis. While cerebrospinal fluid (CSF) data were unavailable, the clinical constellation aligns with previous descriptions of dengue-associated encephalopathy and post-infectious neuroinflammation. DENV-3 has previously been shown to be one of the dengue virus types with the highest neurotropism, especially certain variants [44,45]. In the region where the present study was conducted, DENV-1, -2, and -3 were circulating. It is plausible that DENV-3 in this area may exhibit greater neurovirulence; however, the possibility that DENV-1 or DENV-2 may be responsible cannot be ruled out [46], highlighting the need for further specific investigation.

Polyserositis (presence of pleural effusion and/or ascites) was observed in all patients who died, consistent with capillary leak syndrome. However, these findings occurred in the absence of classical hemoconcentration, as indicated by lower hematocrit levels in non-survivors. This discrepancy points to a severe hypoalbuminemic state, as confirmed by significantly reduced serum albumin levels (median 2.00 g/dL), which impairs intravascular oncotic pressure and promotes third-spacing. Capillary leakage without hemoconcentration suggests an atypical presentation, possibly involving a more aggressive or dysregulated inflammatory response, or it may reflect a secondary effect of treatment—particularly the administration of intravenous fluids [47,48]. Hypoalbuminemia may also reflect malnutrition, systemic capillary leak, or hepatic dysfunction [49,50]. Interestingly, AST and ALT levels were lower in non-survivors than in survivors, which may suggest that liver involvement was not a determining factor in the fatal outcomes of these patients, contrary to what has been reported in previous studies [43].

Among all biochemical parameters, BUN (AUC 0.91) and urea (AUC 0.89) demonstrated the highest discriminative power for mortality. Creatinine levels were also elevated, albeit with a smaller magnitude. These findings suggest acute kidney injury (AKI) as a central contributor to dengue mortality in this cohort, despite the absence of pre-existing chronic kidney disease. Renal impairment in dengue has often been considered a terminal event or consequence of severe plasma leakage [44–46]. However, emerging evidence suggests that AKI may develop early due to direct viral injury, systemic inflammation, or renal hypoperfusion [44–46]. Moreover, uremia may contribute to encephalopathy and seizures, as discussed earlier, forming a vicious cycle. Although MPV was not significantly different between groups in our cohort, we included it due to prior studies suggesting its relevance in inflammatory and infectious conditions, including dengue. Its inclusion may guide future analyses of hematologic indicators associated with disease severity [51].

High RDW values in fatal cases (13.2% vs. 12.7%) may indicate ineffective erythropoiesis or oxidative stress-induced hemolysis, both linked to systemic inflammation. An increased red cell distribution width (RDW) has been linked to higher levels of pro-inflammatory cytokines, including tumor necrosis factor alpha and interleukin-6. These inflammatory mediators reduce the effectiveness of erythropoietin and promote the release of immature red blood cells into the circulation, which contributes to the elevation of RDW [47,48]. While these markers are non-specific, their consistent elevation in the fatal group suggests that dengue-related deaths may result from an uncontrolled immune cascade, as has been reported in other diseases such as COVID-19 [49], rather than isolated organ failure. All deceased patients received paracetamol, and half had comorbid diabetes and hypertension, with increased use of antihypertensives. While NSAIDs and steroids were not administered (appropriately, per guidelines), the high burden of systemic inflammation observed calls into question whether anti-inflammatory adjuncts, such as corticosteroids or IL-6 inhibitors, might be beneficial in a subset of patients with hyper-inflammatory profiles. A previous study by Espinoza-Gómez (2017) [50] suggested that the administration of methylprednisolone may be beneficial in managing patients with dengue presenting warning signs [50]. However, a systematic review of randomized trials in dengue has so far shown no clear benefit from the use of corticosteroids [52].

This study has certain limitations that should be acknowledged. First, it was conducted at a single referral center in Colima, Mexico, which may limit the generalizability of the findings to other regions with different circulating DENV serotypes or healthcare settings. Second, although several clinically relevant differences were identified between survivors and non-survivors, the limited number of fatal outcomes available for analysis constrains the statistical power for some comparisons. To address this, we applied appropriate non-parametric statistical methods to ensure the robustness of our results. Nonetheless, expanding the study population through multicenter collaborations or incorporating data from other regions could allow for a broader representation of disease severity, improve the external validity of the findings, or enable the detection of other potential biomarkers. It is important to acknowledge that the clinical heterogeneity observed may also be influenced by previous exposures to other dengue virus serotypes which were not assessed in this study. The phenomenon of immune imprinting or antibody-dependent enhancement (ADE) could contribute to the severity of disease manifestations, especially in secondary infections, and may partly explain the atypical laboratory and clinical profiles observed in our non-survivor group [53]. Incorporating serological or immunological data in future research will be essential to disentangle these effects.

Third, its retrospective design restricts the ability to infer causality, and certain advanced diagnostics—such as cerebrospinal fluid analysis, viral genotyping, or imaging studies—were not uniformly available. These tools, while not essential for the initial objectives of this study, represent important opportunities for future research aiming to further elucidate the mechanisms underlying neurological and renal involvement.

Finally, the absence of cytokine profiling and other inflammatory markers limits immunopathological insight yet highlights the need to incorporate these variables in subsequent, more comprehensive studies. Despite these limitations, the findings contribute valuable clinical observations that can inform early recognition and management strategies in severe dengue cases.

This study is based on a single-center retrospective study with a small number of fatal outcomes. Despite this, the consistent clinical and biochemical patterns observed among non-survivors highlight biologically plausible and clinically relevant phenomena. Prospective validation in larger, multicenter cohorts is essential. Furthermore, viral genotyping, cytokine profiling, and imaging studies (e.g., brain MRI, renal ultrasound) could clarify the mechanisms of neuro-renal involvement. Finally, updated clinical scores incorporating BUN, seizures, and albumin levels may provide more accurate triage tools during DENV-3 outbreaks, particularly when classical warning signs (like thrombocytopenia) are absent. The need for mechanical ventilation in all fatal cases also underscores the importance of early recognition and referral. In resource-limited settings, delays in critical care access may worsen outcomes, particularly for neurological and respiratory complications.

5. Conclusions

In conclusion, the 2024 DENV outbreak in western Mexico (with increasing presence of DENV-3) revealed a lethal clinical phenotype characterized by neurological involvement, polyserositis, systemic inflammation, and early renal dysfunction, in the absence of severe thrombocytopenia. These findings challenge traditional assumptions about dengue mortality and underscore the heterogeneity of dengue pathophysiology across serotypes and populations. Timely recognition of atypical patterns, especially in older adults with comorbidities, is critical to reducing dengue-related deaths. A paradigm shift is urgently needed, both in clinical training and in WHO severity classifications, to include non-hemorrhagic markers of poor prognosis in dengue.

Author Contributions: Conceptualization, I.D.-E.; data curation, E.G.-S. and F.E.-G.; formal analysis, M.A.M.-H., P.G.d.L.-F. and V.M.; funding acquisition, I.D.-E.; investigation, J.D.-M., F.R.-L., M.Á.-G., J.G.-E. and V.M.; methodology, M.A.M.-H., J.D.-M., G.A.H.-F., F.R.-L., K.A.C.-C., P.G.d.L.-F., D.A.R.-C., R.A.-C., E.G.-S., M.Á.-G., J.G.-E., V.M. and F.E.-G.; project administration, I.D.-E.; software, M.A.M.-H., K.A.C.-C., D.A.R.-C. and R.A.-C.; supervision, I.D.-E.; validation, M.A.M.-H., J.D.-M., G.A.H.-F., F.R.-L., K.A.C.-C., P.G.d.L.-F., D.A.R.-C., R.A.-C., E.G.-S. and F.E.-G.; visualization, G.A.H.-F.; writing—original draft, G.A.H.-F.; writing—review and editing, G.A.H.-F. and I.D.-E. All authors have read and agreed to the published version of the manuscript.

Funding: This research received no external funding.

Institutional Review Board Statement: This study was conducted at the Regional University Hospital of Colima "Hospital Regional Universitario de Colima", as part of a project that analyzes the epidemiology of dengue over time. It was approved by the local research ethics committee (approval code: CI 2024/2/SR/MI/254). This study adhered to the Strengthening the Reporting of Observational Studies in Epidemiology (STROBE) guidelines. As it was a retrospective analysis based on medical records, the ethics committee waived the requirement for informed consent. Nonetheless, strict measures were taken to ensure the anonymity and confidentiality of all patient data.

Informed Consent Statement: As it was a retrospective analysis based on medical records, the ethics committee waived the requirement for informed consent. Nonetheless, strict measures were taken to ensure the anonymity and confidentiality of all patient data.

Data Availability Statement: The original contributions presented in this study are included in the article; further inquiries can be directed to the corresponding author.

Acknowledgments: The authors would like to thank Julio V. Barrios Nuñez from ICEP Colima, Mexico, for his assistance with editing the English language. G.A. Hernández-Fuentes gratefully acknowledges the financial support provided by SECIHTI, Mexico, for his postdoctoral studies (Grant No. 633738).

Conflicts of Interest: The authors declare no conflicts of interest.

References

1. Dengue and Severe Dengue. Available online: https://www.who.int/news-room/fact-sheets/detail/dengue-and-severe-dengue (accessed on 2 June 2025).
2. Jing, Q.; Wang, M. Dengue Epidemiology. *Glob. Health J.* **2019**, *3*, 37–45. [CrossRef]
3. Pan American Health Organization; WHO. In Record Year of Dengue Cases, PAHO Urges Countries to Strengthen Response as Seasonal Transmission Set to Begin in South America. Available online: https://www.paho.org/en/news/8-10-2024-record-year-dengue-cases-paho-urges-countries-strengthen-response-seasonal (accessed on 2 June 2025).
4. Secretaría de Salud, Gobierno de Mexico. Panorama Epidemiológico de Dengue 2024. Available online: https://www.gob.mx/salud/documentos/panorama-epidemiologico-de-dengue-2024 (accessed on 3 June 2025).
5. Rodrigo, C.; Sigera, C.; Fernando, D.; Rajapakse, S. Plasma Leakage in Dengue: A Systematic Review of Prospective Observational Studies. *BMC Infect. Dis.* **2021**, *21*, 1082. [CrossRef]
6. Kothari, D.; Patel, N.; Bishoyi, A.K. Dengue: Epidemiology, Diagnosis Methods, Treatment Options, and Prevention Strategies. *Arch. Virol.* **2025**, *170*, 48. [CrossRef]
7. Horstick, O.; Martinez, E.; Guzman, M.G.; San Martin, J.L.; Ranzinger, S.R. WHO Dengue Case Classification 2009 and Its Usefulness in Practice: An Expert Consensus in the Americas. *Pathog. Glob. Health* **2015**, *109*, 19–25. [CrossRef]
8. Del Carpio-Orantes, L.; Trelles-Hernández, D.; López-Vargas, E.R.; Munguía-Sereno, Á.E. Atypical Presentations of Denguevirus 3 in Veracruz, Mexico. *Travel. Med. Infect. Dis.* **2023**, *56*, 102657. [CrossRef]
9. Organización Panamericana de La Salud. La OPS Alerta Sobre El Riesgo de Brotes de Dengue Por La Circulación Del Serotipo DENV-3 En Las Américas-OPS/OMS. Available online: https://www.paho.org/es/noticias/10-2-2025-ops-alerta-sobre-riesgo-brotes-dengue-por-circulacion-serotipo-denv-3-americas (accessed on 3 June 2025).
10. Tatura, S.N.N.; Denis, D.; Santoso, M.S.; Hayati, R.F.; Kepel, B.J.; Yohan, B.; Sasmono, R.T. Outbreak of Severe Dengue Associated with DENV-3 in the City of Manado, North Sulawesi, Indonesia. *Int. J. Infect. Dis.* **2021**, *106*, 185–196. [CrossRef]
11. Wilder-Smith, A.; Gubler, D.J.; Weaver, S.C.; Monath, T.P.; Heymann, D.L.; Scott, T.W. Epidemic Arboviral Diseases: Priorities for Research and Public Health. *Lancet Infect. Dis.* **2017**, *17*, e101–e106. [CrossRef]
12. Sauceda-Acosta, D.; Almendares, S.P.P.; Cárcamo, E.; Zúniga-Gutiérrez, M.; Beltrán, B.; Rivera, M.F.; Rodríguez, M.M.; Enamorado, J. Risk Factors for Dengue Mortality: A 7-Year Retrospective Cohort in Honduras. *BMC Infect. Dis.* **2025**, *25*, 1–12. [CrossRef]
13. Mendoza-Cano, E.F.; Lugo-Radillo, O.; Ortega-Ramírez, A.; Murillo-Zamora, A.D.; Ríos-Bracamontes, F.; Mendoza-Cano, O.; Lugo-Radillo, A.; Daniela Ortega-Ramírez, A.; Murillo-Zamora, E. Factors Contributing to In-Hospital Mortality in Dengue: Insights from National Surveillance Data in Mexico (2020–2024). *Trop. Med. Infect. Dis.* **2024**, *9*, 202. [CrossRef]
14. Liu, L.T.; Huang, S.Y.; Lin, C.H.; Chen, C.H.; Tsai, C.Y.; Lin, P.C.; Tsai, J.J. The Epidemiology and Identification of Risk Factors Associated with Severe Dengue during the 2023 Dengue Outbreak in Kaohsiung City, Taiwan. *Travel. Med. Infect. Dis.* **2025**, *65*, 102852. [CrossRef]
15. von Elm, E.; Altman, D.G.; Egger, M.; Pocock, S.J.; Gøtzsche, P.C.; Vandenbroucke, J.P. The Strengthening the Reporting of Observational Studies in Epidemiology (STROBE) Statement: Guidelines for Reporting Observational Studies. *J. Clin. Epidemiol.* **2008**, *61*, 344–349. [CrossRef]
16. Zandecki, M.; Genevieve, F.; Gerard, J.; Godon, A. Spurious Counts and Spurious Results on Haematology Analysers: A Review Part II: White Blood Cells, Red Blood Cells, Haemoglobin, Red Cell Indices and Reticulocytes. *Int. J. Lab. Hematol.* **2007**, *29*, 21–41 [CrossRef]
17. Brihi, J.E.; Pathak, S. Normal and Abnormal Complete Blood Count with Differential. In *StatPearls*; StatPearls Publishing: Treasure Island, FL, USA, 2024.
18. Mandrekar, J.N. Receiver Operating Characteristic Curve in Diagnostic Test Assessment. *J. Thorac. Oncol.* **2010**, *5*, 1315–1316 [CrossRef]
19. Zweig, M.H.; Campbell, G. Receiver-Operating Characteristic (ROC) Plots: A Fundamental Evaluation Tool in Clinical Medicine *Clin. Chem.* **1993**, *39*, 561–577. [CrossRef]
20. Ferdinand, R.F. Validity of the CBCL/YSR DSM-IV Scales Anxiety Problems and Affective Problems. *J. Anxiety Disord.* **2008**, *22* 126–134. [CrossRef]
21. Rosner, B. *Fundamentals of Biostatistics*, 7th ed.; Brooks/Cole, Cengage Learning: Boston, MA, USA, 2011; Volume 1.

22. Dudley, W.N.; Benuzillo, J.G.; Carrico, M.S. SPSS and SAS Programming for the Testing of Mediation Models. *Nurs. Res.* **2004**, *53*, 59–62. [CrossRef]
23. del año, A.; Fajardo-Dolci, G.; Meljem-Moctezuma, J.; Vicente-González, E.; Vicente Venegas-Páez, F.; Villalba-Espinoza, I.; Luisa Pérez-Cardoso, A.; Adrián Barrón-Saldaña, D.; Barragán-Ramírez, C.; Novoa-Boldo, A.; et al. Analysis of Dengue Fever Deaths in Mexico: 2009. *Rev. Med. Inst. Mex. Seguro Soc.* **2012**, *50*, 589–598.
24. Ilic, I.; Ilic, M. Global Patterns of Trends in Incidence and Mortality of Dengue, 1990–2019: An Analysis Based on the Global Burden of Disease Study. *Medicina* **2024**, *60*, 425. [CrossRef]
25. Moraes, G.H.; Duarte, E.D.F.; Duarte, E.C. Determinants of Mortality from Severe Dengue in Brazil: A Population-Based Case-Control Study. *Am. J. Trop. Med. Hyg.* **2013**, *88*, 670. [CrossRef]
26. Teixeira, O.F.B.; Xavier, S.P.L.; de Carvalho Félix, N.D.; da Silva, J.W.M.; de Abreu, R.M.S.X.; Miranda, K.C.L. Repercusiones de La Pandemia de COVID-19 Para Las Personas Con Autismo y Sus Familias: Revisión de Alcance. *Rev. Lat. Am. Enferm.* **2022**, *30*, e3728. [CrossRef]
27. Bisanzio, D.; Estofolete, C.F.; Reithinger, R. Dengue and Diabetes Comorbidity: An Emerging Public Health Threat. *Int. Health* **2025**, *17*, 597–599. [CrossRef]
28. Sekaran, S.D.; Liew, Z.M.; Yam, H.C.; Raju, C.S. The Association between Diabetes and Obesity with Dengue Infections. *Diabetol. Metab. Syndr.* **2022**, *14*, 1–12. [CrossRef]
29. Shawon, S.R.; Hamid, M.K.I.; Ahmed, H.; Khan, S.A.; Dewan, S.M.R. Dengue Fever in Hyperglycemic Patients: An Emerging Public Health Concern Demanding Eyes on the Effective Management Strategies. *Health Sci. Rep.* **2024**, *7*, e70144. [CrossRef]
30. Zhu, T.; Xiao, X.; Zhu, X.; Wang, X. Hospitalised Dengue Patients and Risk of Hypertension: A Systematic Review and Meta-Analysis. *Rev. Med. Virol.* **2025**, *35*, e70013. [CrossRef]
31. De Azeredo, E.L.; Monteiro, R.Q.; De-Oliveira Pinto, L.M. Thrombocytopenia in Dengue: Interrelationship between Virus and the Imbalance between Coagulation and Fibrinolysis and Inflammatory Mediators. *Mediat. Inflamm.* **2015**, *2015*, 313842. [CrossRef]
32. de Souza Andrade, A.; Oliveira Campos, S.; Dias, J.; Campos, M.A.; Kroon, E.G. Dengue Virus 3 Genotype I (GI) Lineage 1 (L1) Isolates Elicit Differential Cytopathic Effect with Syncytium Formation in Human Glioblastoma Cells (U251). *Virol. J.* **2023**, *20*, 204. [CrossRef]
33. Phu Ly, M.H.; Takamatsu, Y.; Nabeshima, T.; Pham Hoai, L.L.; Pham Thi, H.; Dang Thi, D.; Nguyen, N.L.; Nguyen Thi, T.T.; Le Thi, Q.M.; Buerano, C.C.; et al. Isolation of Dengue Serotype 3 Virus from the Cerebrospinal Fluid of an Encephalitis Patient in Hai Phong, Vietnam in 2013. *J. Clin. Virol.* **2015**, *70*, 93–96. [CrossRef]
34. Leng, X.; Yang, H.; Zhao, L.; Feng, J.; Jin, K.; Liao, L.; Zhang, F. Dengue Encephalopathy in an Adult Due to Dengue Virus Type 1 Infection. *BMC Infect. Dis.* **2024**, *24*, 1–4. [CrossRef]
35. Patel, J.P.; Saiyed, F.; Hardaswani, D. Dengue Fever Accompanied by Neurological Manifestations: Challenges and Treatment. *Cureus* **2024**, *16*, e60961. [CrossRef]
36. Castellanos, J.; Bello, J.; Velandia-Romero, M. Neurological Manifestations during Dengue Virus Infection. *Infectio* **2014**, *18*, 167–176. [CrossRef]
37. Kulkarni, R.; Pujari, S.; Gupta, D. Neurological Manifestations of Dengue Fever. *Ann. Indian Acad. Neurol.* **2021**, *24*, 693–702. [CrossRef]
38. Dhenni, R.; Karyanti, M.R.; Putri, N.D.; Yohan, B.; Yudhaputri, F.A.; Ma'roef, C.N.; Fadhilah, A.; Perkasa, A.; Restuadi, R.; Trimarsanto, H.; et al. Isolation and Complete Genome Analysis of Neurotropic Dengue Virus Serotype 3 from the Cerebrospinal Fluid of an Encephalitis Patient. *PLoS Negl. Trop. Dis.* **2018**, *12*, e0006198. [CrossRef]
39. Phu Ly, M.H.; Nguyen, C.T.; Nguyen, T.V.; Ngan Nguyen, T.T.; Nabeshima, T.; Adungo, F.; Takamatsu, Y.; Huy, N.T.; Mai Le, T.Q.; Morita, K.; et al. Differential Infectivity of Human Neural Cell Lines by a Dengue Virus Serotype-3 Genotype-III with a Distinct Nonstructural Protein 2A (NS2A) Amino Acid Substitution Isolated from the Cerebrospinal Fluid of a Dengue Encephalitis Patient. *Can. J. Infect. Dis. Med. Microbiol.* **2023**, *2023*, 1–11. [CrossRef]
40. Siddall, E.; Radhakrishnan, J. Capillary Leak Syndrome: A Cytokine and Catecholamine Storm? *Kidney Int.* **2019**, *95*, 1009–1011. [CrossRef]
41. Agrawal, S.; Kumar, S.; Talwar, D.; Patel, M.; Reddy, H. Significance of Neutrophil-Lymphocyte Ratio, Neutrophil-Platelet Ratio, and Neutrophil-Tolymphocyte and Platelet Ratio in Predicting Outcomes in Dengue Patients on Admission in Wardha, Maharashtra, India: A Retrospective Cohort Study. *J. Clin. Diagn. Res.* **2023**, *12*, OC1–OC4. [CrossRef]
42. Mendoza-Hernandez, M.A.; Guzman-Esquivel, J.; Ramos-Rojas, M.A.; Santillan-Luna, V.V.; Sanchez-Ramirez, C.A.; Hernandez-Fuentes, G.A.; Diaz-Martinez, J.; Melnikov, V.; Rojas-Larios, F.; Martinez-Fierro, M.L.; et al. Differences in the Evolution of Clinical, Biochemical, and Hematological Indicators in Hospitalized Patients with COVID-19 According to Their Vaccination Scheme: A Cohort Study in One of the World's Highest Hospital Mortality Populations. *Vaccines* **2024**, *12*, 72. [CrossRef]
43. Campana, V.; Inizan, C.; Pommier, J.D.; Menudier, L.Y.; Vincent, M.; Lecuit, M.; De Lamballerie, X.; Dupont-Rouzeyrol, M.; Murgue, B.; Cabié, A. Liver Involvement in Dengue: A Systematic Review. *Rev. Med. Virol.* **2024**, *34*, e2564. [CrossRef]

44. Awad, A.A.; Khatib, M.N.; Gaidhane, A.M.; Ballal, S.; Bansal, P.; Srivastava, M.; Arora, I.; Kumar, M.R.; Sinha, A.; Rawat, P.; et al Predictors of Acute Kidney Injury in Dengue Patients: A Systematic Review and Meta-Analysis. *Virol. J.* **2024**, *21*, 223. [CrossRef]
45. Bignardi, P.R.; Pinto, G.R.; Boscarioli, M.L.N.; Lima, R.A.A.; Delfino, V.D.A. Acute Kidney Injury Associated with Dengue Virus Infection: A Review. *Braz. J. Nephrol.* **2022**, *42*, 232–237. [CrossRef]
46. Ng, W.Y.; Atan, R.; Yunos, N.M.; Bin Md Kamal, A.H.; Roslan, M.H.; Quah, K.Y.; Teh, K.X.; Zaid, M.; Kassim, M.; Mariapun J.; et al. A Double Whammy: The Association between Comorbidities and Severe Dengue among Adult Patients—A Matched Case-Control Study. *PLoS ONE* **2022**, *17*, e0273071. [CrossRef]
47. Förhécz, Z.; Gombos, T.; Borgulya, G.; Pozsonyi, Z.; Prohászka, Z.; Jánoskuti, L. Red Cell Distribution Width in Heart Failure Prediction of Clinical Events and Relationship with Markers of Ineffective Erythropoiesis, Inflammation, Renal Function, and Nutritional State. *Am. Heart J.* **2009**, *158*, 659–666. [CrossRef]
48. Rosas-Cabral, A.; Prieto-Macías, J.; Gutiérrez-Padilla, P.; Autónoma De Aguascalientes, U. Asociación Entre La Anchura de La Distribución Del Eritrocito y El Síndrome Metabólico. *Lux Médica* **2023**, *18*, 52.
49. Bommenahalli Gowda, S.; Gosavi, S.; Ananda Rao, A.; Shastry, S.; Raj, S.C.; Menon, S.; Suresh, A.; Sharma, A. Prognosis of COVID-19: Red Cell Distribution Width, Platelet Distribution Width, and C-Reactive Protein. *Cureus* **2021**, *13*, e13078. [CrossRef]
50. Espinoza- Gómez, F.; Aréchiga Ramírez, J.C.; Sánchez Gómez, J.H.; Newton Sánchez, O.A.; Delgado Enciso, I.; Lopez Lemus A.U.; Rojas-Larios, F. Effect of the Methylprednisolone in the Progression of Dengue with Warning Signs. a Clinical Trial. *Int. J Curr. Res.* **2017**, *9*, 59368–59372.
51. Tajarernmuang, P.; Phrommintikul, A.; Limsukon, A.; Pothirat, C.; Chittawatanarat, K. The Role of Mean Platelet Volume as a Predictor of Mortality in Critically Ill Patients: A Systematic Review and Meta-Analysis. *Crit. Care Res. Pract.* **2016**, *2016*, 4370834 [CrossRef]
52. Panpanich, R.; Sornchai, P.; Kanjanaratanakorn, K. Corticosteroids for Treating Dengue Shock Syndrome. *Cochrane Database Syst Rev.* **2006**, *19*, CD003488. [CrossRef]
53. Aynekulu Mersha, D.G.; van der Sterren, I.; van Leeuwen, L.P.M.; Langerak, T.; Hakim, M.S.; Martina, B.; van Lelyveld, S.F.L van Gorp, E.C.M. The Role of Antibody-Dependent Enhancement in Dengue Vaccination. *Trop. Dis. Travel. Med. Vaccines* **2024** *10*, 22. [CrossRef]

viruses

MDPI

Article

Lineage B Genotype III of Dengue Virus Serotype 3 (DENV-3III_B) Is Responsible for Dengue Outbreak in Dire Dawa City, Ethiopia, 2023

Abebe Aseffa Negeri [1,*,†], Dawit Hailu Alemayehu [2,†], Saro Abdella Abrahim [1], Tsigereda Kifle Wolde [1], Gutema Bulti Tura [1], Alemnesh Hailemariam Bedasso [1], Danile Tsega Geretsion [1], Ebise Abose Djirata [1], Eyilachew Zenebe Awule [3], Diana Rojas-Gallardo [4], Asefa Konde Korkiso [1], Kalkidan Melaku [2], Raffael Joseph [2], Abaysew Ayele [2], Mesfin Mengesha Tsegaye [2], Anne Piantadosi [5], Getachew Tollera [1], Alemseged Abdissa [2], Mesay Hailu Dangiso [1], Adane Mihret [2], Andargachew Mulu [2] and Tesfaye Gelanew [2,*]

1 Ethiopian Public Health Institute, Addis Ababa P.O. Box 1242, Ethiopia
2 Armauer Hansen Research Institute, Addis Ababa P.O. Box 1005, Ethiopia
3 Dire Dawa City Administration Health Bureau, Dire Dawa P.O. Box 1377, Ethiopia
4 Population Biology, Ecology and Evolution Graduate Program, Emory University, Atlanta, GA 30322, USA
5 Department of Pathology and Laboratory Medicine, Emory University School of Medicine, Atlanta, GA 30322, USA
* Correspondence: abebea84@gmail.com (A.A.N.); tesfaye.gelanew@ahri.gov.et or tesfayegtaye@gamail.com (T.G.); Tel.: +25-198-015-3960 (T.G.)
† These authors contributed equally to this work.

Abstract: The eastern parts of Ethiopia, including Dire Dawa City, have experienced annual dengue fever (DF) outbreaks since 2013, leading to significant healthcare and economic impacts. However, comprehensive evidence on the specific dengue virus (DENV) serotypes and genotypes involved remains limited. During the 2023 DF outbreak, the National Arbovirus Laboratory received seventy serum samples from suspected DF patients. Positive samples underwent sequencing of the CprM region of the DENV genome, and the obtained sequences were analyzed phylogenetically. Among the patients, 32 (45.7%) displayed early warning signs of severe dengue, and 13 were hospitalized, most showing symptoms indicative of severe dengue. Out of 67 adequate samples, 44 (65.6%) tested positive for DENV RNA by RT-PCR, and 17 successfully underwent CprM sequencing. All sequenced samples were identified as DENV-3, genotype III, major lineage B (DENV-3III_B), with two distinct minor lineages (DENV-3III_B.2 and DENV-3III_B.3). Phylogenetic analysis showed that these lineages were closely related to sequences from the Afar region, suggesting interconnected outbreaks with multiple co-circulating lineages. This study identifies DENV-3III_B as the cause of the 2023 DF outbreak in Dire Dawa City and highlights the need for enhanced viral genomic surveillance in Africa.

Keywords: DENV; DF; serotype; genotype; CprM; DENV-3/GIII; Dire Dawa City; Ethiopia

1. Introduction

Dengue fever (DF), caused by four genetically distinct serotypes of dengue virus (DENV-1 to DENV-4), has become a significant global health challenge. Over the past two decades, the incidence of dengue has increased tenfold, rising from 500,000 reported cases worldwide in 2000 to a staggering 5.2 million cases in 2019. In 2019, dengue outbreaks spread across 129 countries, marking an unprecedented peak. Although there was a slight decline in cases between 2020 and 2022 due to the COVID-19 pandemic and reduced reporting rates, 2023 witnessed a resurgence in dengue cases globally [1–3].

In areas where DF cases have been identified as endemic and/or epidemic, especially in tropical and subtropical countries, nonspecific febrile illnesses are common. Febrile illnesses can be caused by a variety of infectious agents such as malaria parasites, alphaviruses, and flaviviruses, complicating surveillance and response programs for outbreaks and endemic diseases [4]. In addition, the Aedes mosquito species, particularly *Ae. aegypti* and *Ae. albopictus* that transmit DENV, is also responsible for the transmission of other arboviruses (e.g., Zika virus, Yellow Fever virus, and Chikungunya virus), and is common in tropical and subtropical countries [5].

In Ethiopia, the first laboratory-confirmed outbreak of DF occurred in Deri Dawa City in 2013. Since then, Ethiopia has experienced nearly annual outbreaks of DF in multiple regions of the country (Figure 1). In the Afar region, DF outbreaks occurred in 2019 in the Gewane District [6]. In the Ethiopian Somalia region, the first outbreak was reported in Godey town between 2014 and 2015 followed by outbreaks that occurred in Kabridahar in 2017 [7] and Warder Woreda in 2021 [8]. These recurrent outbreaks have impacted an already fragmented health system [9] and the economy of these parts of the country.

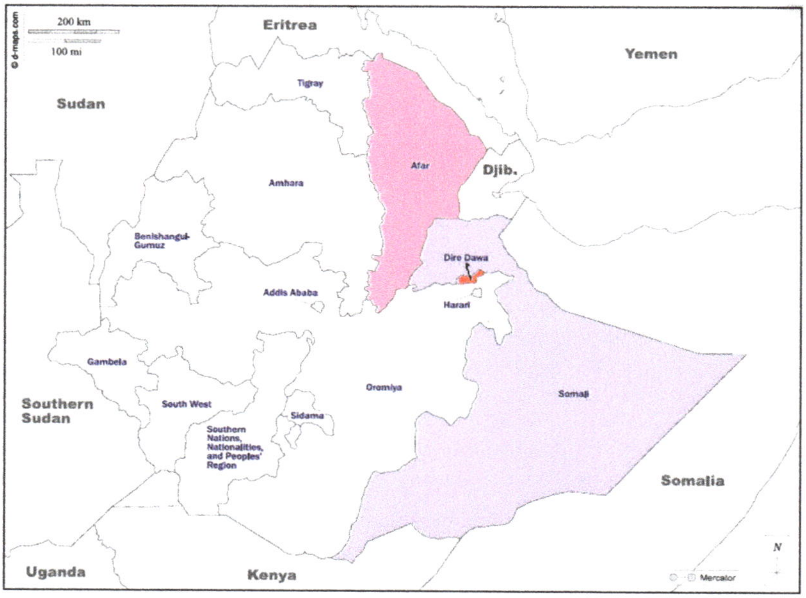

Figure 1. Map showing the Dire Dawa City Administration where the Dengue outbreak occurred, and the study of DENV strains originated. The shaded areas indicate the epicenters of dengue outbreaks in Ethiopia since 2013: Dire Dawa (red), Afar (red-violet), and Somalia (light blue).

In April 2023, an outbreak of DF was initially reported in the Afar region, the northeastern part of Ethiopia. The first affected districts were Logia and Mille. Since April 4, 2023, however, it has spread to all seven districts and towns in the Afar region (Figure 1). As of 26 June 2023, a total of 6133 suspected and confirmed cases were reported, resulting in nine associated deaths (with a case/fatality ratio of 0.5%) [10]. It was documented that 8 of the 10 epidemic samples examined by reverse transcriptase polymerase chain reaction (RT-PCR) triplex [for Dengue, Zika, and Chikungunya] belonged to serotype DENV-3 [10].

As a suspected expansion of the outbreak in the Afar region, a DF outbreak was recorded in Dire Dawa City in December 2023 (Figure 1). However, there are no data regarding the DENV serotypes and genotypes associated with the outbreak. Whether the outbreak was caused by DENV-3 as in the Afar region remains unknown. In the present study, we present data on the serotypes, genotypes, and lineages of the DENV strains responsible for the outbreak in Dire Dawa City during 2023 and discuss the epidemiological and clinical consequences.

2. Materials and Methods

2.1. Ethical Considerations

As our study aimed to better understand DENV strains linked to DF outbreaks, we obtained an ethical approval waiver (protocol number: PO-036-24) from the All-African Leprosy Rehabilitation and Training Center/Armauer Hansen Research (ALERT/AHRI) Ethics Committee. Additionally, we obtained permission from the Dire Dawa City Health Bureau and health facilities' authorities to investigate these outbreak samples.

2.2. Study Settings and Patient Data

On 14 December 2023, the Ethiopia Public Health Institute (EPH) received 70 serum samples from an outbreak of DF in Dire Dawa City Administration (Figure 1 and Table 1). These samples were collected from DF-suspected cases and shipped to the National Arbovirus Laboratory of EPHI for laboratory confirmation of the causative agent as part of an outbreak response investigation (Figure 2). The reported clinical signs and symptoms for each suspected case were extracted from medical charts available at the health facilities they were admitted to, using a case-based surveillance request form. Mild symptoms include sudden onset of high fever, vomiting, nausea, rash, arthralgia (joint pain), headache, chills, myalgia (muscle pain), sore throat, dizziness, runny nose, pruritus (itchiness), adenosis, diarrhea, and shivering. Warning signs indicative of severe dengue include belly pain or tenderness, persistent vomiting (at least 3 times/24 h), bleeding from the nose or gum, being restless or irritable, feeling weak, conjunctiva congestion, and convulsions (Table 2). The categories of mild symptoms and warning signs indicative of severe dengue are based on the WHO severe dengue classification [11].

Table 1. Status of admission and demographic characteristics of dengue fever suspected cases (n = 70) reported at health facilities in the Dire Dawa City Administration, 2023.

Variable	Specific Variable	N (%)
Age	Median age 30; IQR [1] 17	
Sex	Male	32 (46) [2]
	Female	38 (54)
Patient admission status	Outpatient	57 (80)
	Inpatient	13 (20)

[1] IQR = Interquartile range. [2] Values in parentheses indicate percentiles (%), while values before parentheses indicate the number (N) of participants.

Table 2. Clinical characteristics of dengue fever suspected cases (n = 70) reported at health facilities in the Dire Dawa City Administration, 2023.

Clinical Symptoms [1]		Yes [2]	No
A. Mild symptoms *	Fever: sudden onset of high fever	66 (98.15%)	1 (1.49%)
	Chills	47 (70.17%)	20 (29.85%)
	Vomiting or nausea	31 (46.27%)	36 (53.73%)

Table 2. *Cont.*

Clinical Symptoms [1]		Yes [2]	No
A. **Mild symptoms** *	Aches and pains: Typically, behind the eyes, muscle, joint, or bone pain	31 (46.27%)	36 (53.73%)
	Myalgia	28 (45.9%)	33 (54.1%)
	Running nose	21 (31.34%)	46 (68.66%)
	Diarrhea	13 (19.4%)	54 (89.60%)
	Rash	8 (13.11%)	53 (86.89%)
	Sore throat	6 (8.89%)	61 (91.04%)
	Dizzy	5 (7.46%)	62 (92.54%)
	Shivering	4 (5.97%)	63 (94.03%)
	Pruritus	2 (2.99%)	65 (97.01%)
	Adenoids	1 (1.52%)	65 (98.48%)
B. **Warning signs indicative of severe dengue**	Weak (Feeling tired)	26 (38.81%)	41 (61.19%)
	Restless or irritable	15 (22.39%)	52 (77.61%)
	Arthralgia	9 (13.43%)	58 (86.57%)
	Persistent vomiting (at least 3 times/24 h)	9 (13.43%)	58 (86.57%)
	Bleeding from the nose or gums	6 (8.96%)	61 (91.04%)
	Belly pain or tenderness	5 (7.46%)	62 (92.54%)
	Conjunctival congestion	3 (4.48%)	64 (95.52%)

[1] Categorization of mild symptoms and warning signs indicative of severe dengue are based on WHO classification. [2] Clinical data for 4 cases out of the 70 suspected dengue fever cases were missing or incomplete. * Mild symptoms indicate a lower level of severity that does not pose an immediate threat to health. Conversely, warning signs indicate a severe dengue that could necessitate emergency intervention.

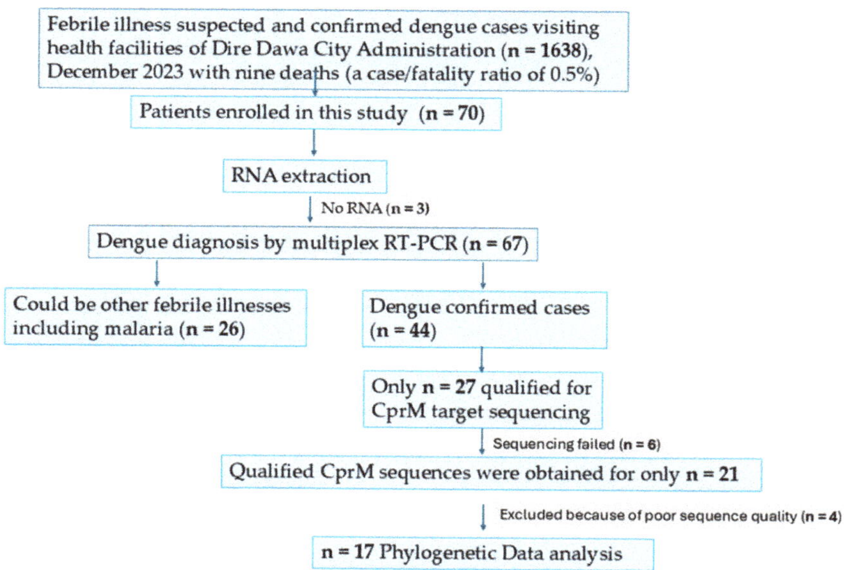

Figure 2. Flow chart showing enrollment of suspected dengue and/or other febrile illnesses, including malaria patients, and molecular detection and characterization of strains isolated during the 2023 dengue outbreak in Dire Dawa, Ethiopia.

2.3. Blood Sample Collection and Processing

Each of the 70 dengue-suspected patients submitted blood samples for laboratory diagnosis of the causative agent for febrile illness. The blood samples were allowed to clot, and the sera were separated. The sera were then shipped in a triple package cold chain to the National Arbovirus Laboratory of the EPHI. At the laboratory, the sera were kept frozen at −20 °C until they were used for DENV viremia detection using reverse transcriptase (RT)- PCR (RT-PCR).

2.4. Viral RNA Extraction and PCR Detection of DENV

Total RNA was isolated from each serum sample using the QIAmp viral RNA kit (QIAGEN, Hilden, Germany). The RNA was eluted in 50 µL of AVE buffer, which was provided with the kit. DENV viremia detection was conducted using RealStar® Dengue RT-PCR kit (Altona Diagnostics, Hamburg-Altona, Germany). Briefly, 20 µL of the master mix (containing primers and probes) was mixed with 10 µL of the RNA sample or control. The mixture was centrifuged for 30 s at approximately 1000× *g* (~3000 rpm). The reactions were incubated for 20 min at 55 °C, followed by 40 cycles of 95 °C for 1 min, 55 °C for 45 s, 72 °C for 15 s, and final incubation at 68 °C for 10 min. The PCR results were interpreted based on the manufacturer's protocol. A threshold cycle (Ct) value of ≤36 for DENV-specific RNA and a Ct value of ≤30 for the Internal Control were considered positive samples for DENV.

2.5. Semi-Nested RT-PCR Amplification (Amplicon Size 511 bp)

RNA samples with a Ct value of 31 or lower and sufficient volume were selected for CprM RT-PCR amplification following the procedure described elsewhere [12]. Briefly, the DENV CprM region (652) bp was first amplified using forward primer (FP): 5′-TCAATATGCTGAAACGCGCGAGAAACCG-3′ (132–153), the reverse primer (RP): 5′-GCGCCTTCNGNNGACATCCA-3′ (764–783), and the high-capacity cDNA reverse transcription kit (Invitrogen, Waltham, MA, USA). The reactions were incubated first at 98 °C for 10 s followed by 35 cycles of 98 °C for 10 s, 60 °C for 20 s, 72 °C for 30 s, and final incubation at 72 °C for 5 min. Semi-nested PCR was also used to amplify 511 bp of the CprM region of the viral genome using the amplified product as a template, and the same FP: 5′-TCAATATGCTGAAACGCGCGAGAAACCG-3′ (132–153) but a different RP-: 5′-TTGCACCAACAGTCAATGTCTTCAGGTTC-3′ (614–642) [12]. The semi-nested amplification was performed in an Applied Biosystems Veriti 96-Well Thermal Cycler (ThermoFisher Scientific, Waltham, MA, USA) using the following cycling parameters: 94 °C for 2 min; 40 cycles of 94 °C for 30 s, 70 °C for 1 s (ramp rate 20%), 55 °C for 45 s (ramp rate 20%), and 65 °C for 3 min, 20 s; and final extension for 10 min at 65 °C [13]. Following the amplification, 5 µL of the first amplicon and semi-nested amplicon size of 652 bp and 511 bp, respectively, was visualized on ethidium bromide-stained agarose gel under UV light. Amplified PCR products (511 bp in size) were then purified using a QIAquick PCR Purification Kit (QIAgen, Germany) and sequenced in both strands using a Big Dye Terminator Cycle Sequencing kit (Applied Biosystems, Waltham, MA, USA) with the same primers that were used for nested PCR. The CprM sequences were confirmed by BLAST. The forward and reverse sequences were aligned and manually edited using Bio Edit software (version 7.2) to obtain the consensus sequence for each sample, resulting in a 405 bp sequence spanning nucleotides 245–649 compared to reference sequence NC_001475. The obtained sequences were analyzed using Dengue Virus Typing Tool at Genome Detective server [13] to identify the serotypes, genotypes, and lineages of DENV associated with the outbreak.

2.6. Phylogenetic Analysis

Although CprM sequences were obtained for 21 DENV strains from Dire Dawa City and were of sufficient quality to determine serotype and genotype, only 16 consensus sequences were used for lineage and sublineage analysis with the specified tool. This was because 5 strains were excluded due to incomplete forward or reverse sequences and poor-quality chromatograms, which made relying on single-strand sequences impractical. The CprM consensus sequences for these 16 DENV strains from Dire Dawa City have been deposited in GenBank under accession numbers PP751832, PP751836-PP751845, and PP7547-PP751852. All DENV-3 CprM sequences available in the NCBI Virus database (https://www.ncbi.nlm.nih.gov/labs/virus/vssi/#/; accessed on 31 January 2025) were downloaded and aligned. Duplicates and identical sequences from the same country and collection date were removed. Representative sequences for DENV-3 genotypes I, II, and V were also included as references. The final data set includes a total of 942 DENV-3 CprM gene sequences collected between 2006 and 2023 across the globe.

For phylogenetic reconstruction, sequences were aligned using MAFFT (version 7.520) [14], and a maximum likelihood tree was constructed with the IQ-TREE software (version 2.2.2.7) [15] using the GTR+F+I+R3 substitution model (according to AICc) with a bootstrap value of 1000. Finally, the resulting tree was visualized and annotated using iTOL (Interactive Tree of Life) [16].

3. Results

3.1. Patient Characteristics

Among the 70 suspected dengue fever (DF) patients, 38 (54%) were female, as indicated in Table 1. The median age was 30 with an interquartile (IQ) range of 17 years. Most (81%) of the suspected cases visited health facilities as outpatients (Table 1), while the remaining 13 cases required inpatient hospitalization for a duration of 2 to 10 days.

The clinical signs/symptoms observed in these (n = 70) DF-suspected cases, whether admitted as outpatients or inpatients, covered a wide range of symptoms consistent with febrile illness. Nearly all of them exhibited three or more clinical symptoms consistent with mild DF. However, regardless of hospitalization status, 32 (45.7%) of these patients displayed one or more early warning signs indicative of severe dengue [11]. Among the 13 patients who were hospitalized for 2 to 10 days, all except two had at least one symptom indicative of severe dengue. The most commonly observed clinical presentations were acute fever (n = 66; 98.15%), headache (n = 65; 97.01%) and chills (n = 47; 70.17%) followed by aches and pains (n = 31; 46.27%) and vomiting/nausea (n = 31; 46.27%) (Table 2). In addition, all of the cases with early warning signs of severe dengue were later confirmed to harbor DENV viremia using RT-PCR.

3.2. DENV Viremia Detection

Among 70 specimens, 3 samples failed to pass the DENV nucleic acid amplification quality control. Among the 67 samples that passed quality control, 65.67% (44/67) had DENV RNA detected (Figure 2).

3.3. DENV Sequence Classifications

Based on rRT-PCR CT < 31, 27 samples out of 44 DENV viremia-positive samples were selected for CprM sequencing, and 21 were successful (Figure 2) allowing classification of their serotypes, genotypes, and lineages using a newly developed lineage system tool (https://dengue-lineages.org) [13].

The sequence quality for five samples was insufficient, and these sequences were excluded from the DENV minor lineage analysis using the new dengue virus typing tool. According to this newly proposed DENV nomenclature [13], the dengue virus typing tool classified all 942 DENV-3 CprM gene sequences collected between 2006 and 2023, including those from our study, as DENV-3II. These sequences were further divided into minor lineages: DENV-3III_B.1, DENV-3III_B.2, and DENV-3III_B.3. All these minor lineages of DENV-3III_B were also found among sequences from the African continent, although some sequences were not assigned to specific minor lineages by the tool. Notably, the new tool assigned all 17 of our sequences to DENV-3, genotype III, major lineage B. Specifically, the minor lineages were assigned as follows: 1 sequence (DENV-3-1357) belonged to minor lineage 3 (DENV-3III_B.3), and 15 sequences (DENV-3-1355, -1363, -1395, -1399, -1400, -1401, -1403, -1404, -1405, -1410, -1411, -1412, -1414, -1415, and -1416) belonged to minor lineage 2 (DENV-3III_B.2). Nonetheless, the new lineage assignment tools did not assign one sequence (DENV-3-1402) to a minor lineage. This sequence was found to be related to minor lineage B2, but not part of it. Although minor lineages 1 (B.1), 2 (B.2), and 3 (B.3) have been documented in DENV 3 lineage III [13], the DENV strains investigated in our study were not assigned to minor lineage 1 (B.1) based on the classification determined by the recently developed dengue virus tool (Figure 3 and Supplementary Figure S1). Supplementary Table S1 presents the 17 DENV strains for which high-quality CprM sequences were obtained and deposited into the GenBank database repository, along with their corresponding accession numbers, serotypes, genotypes, linages, and minor lineages—assigned to them by the newly developed lineage assignment tools [13].

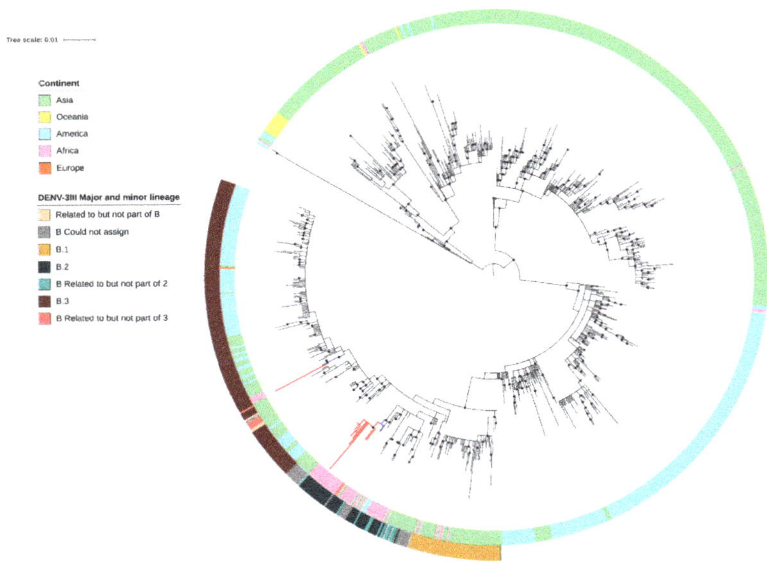

Figure 3. Maximum likelihood phylogenetic tree of DENV-3 genotype III (DENV-3III) strains isolated during the 2023 outbreak in Dire Dawa, Ethiopia. The tree includes CprM sequences newly generated in this study (red branches, N = 17) and representative reference sequences from DENV-3III (N = 942, sequences from Ethiopia in blue and other branches in black). The outer circles indicate continent of origin and the DENV-3III_B major and minor lineages. Nodes with ultrafast bootstrap values of 95 and above are indicated with black circles. The internal tree scale represents the number of nucleotide substitutions per site. A more detailed tree with branch labels, including accession number, country of origin, and year of isolation, as well as the percentages of support for the strains' genotype, major lineage, and minor lineage assignments using recently developed DENV lineage assignment tools, can be found in Supplementary Figure S1 and Supplementary Table S1, respectively.

3.4. Phylogenetic Tree

We investigated the phylogenetic relationships between the 17 new DENV-3III_B CprM sequences obtained in this study and other DENV-3III_B CprM sequences from various regions, including Africa (Kenya, Gabon, Burkina Faso, Mozambique, Tanzania, and Ethiopia), Asia (the Middle East, India, Sri Lanka, Thailand, and China), the Americas (Jamaica, Martinique, Dominican Republic, and the USA), and Europe (Italy), deposited in a public sequence repository (DDBJ/ENA/GenBank-INSDC). As expected, the 15 DENV-3III_B minor lineage 2 (DENV-3III_B.2) sequences, including the one related to B.2, clustered separately from the single DENV-3III_B minor lineage 3 (DENV-3III_B.3) sequence (Figure 3). Both the DENV-3III_B.2 and DENV-3III_B.3 sequences from this study clustered closely with reference sequences isolated from the 2023 outbreak in the Afar region of Ethiopia [17], supporting the hypothesis that the outbreak in Dire Dawa was related to the outbreak in the Afar region, and indicating that multiple lineages co-circulated in both outbreaks.

4. Discussion

Knowledge of the genetic diversity of DENV strains responsible for outbreaks is essential for efficient management, vaccine development, and outbreak prevention. Though Ethiopia's eastern region, including Dire Dawa City, has seen many DF outbreaks since 2013, there is scant information regarding the genetic diversity of circulating DENV strains. In this study, we used genetic analysis of the DENV CprM region to demonstrate that the 2023 outbreak in Dire Dawa City was due to two co-circulating lineages of DENV-3III_B, each of which was closely related to sequences from a related outbreak in the Afar region of Ethiopia in 2023. Our results are consistent with a recent study that analyzed 7 full DENV-3III genome sequences from the Afar outbreak [17], supporting the hypothesis that the outbreak in Dire Dawa was related to the outbreak in the Afar region, and indicating that multiple lineages co-circulated in both the Afar and Dire Dawa outbreaks. However, the paucity of regional DENV sequence data makes it difficult to determine the timing and origin of these introductions. Recently, other DENV-3III outbreaks have been reported in East Africa, notably in South Sudan in 2022 [18] and in Kenya in 2019 [19]. We were not able to assess the genetic relatedness between the current strains and outbreak strains from these neighboring countries due to the fact that they employed other targets for sequencing, such as the E gene. These results underscore the need for further genomic surveillance of DENV in East Africa, ideally using full-length genome sequences.

In contrast to the prior dengue outbreaks in Ethiopia which were mild [7,20–22], in this study, 46% (n = 32) of the 70 DF-suspected cases displayed early warning signs indicative of severe dengue, while the remaining cases showed mild symptoms. All of these cases with early warning signs of severe dengue were confirmed to harbor DENV viremia using RT-PCR.

We propose three scenarios that might have led to the occurrence of severe cases of DF during the 2023 outbreak in Dire Dawa City. One possibility is that the observed severe dengue cases could be secondary DENV-3III infections that happened after primary infection with a different serotype. In support of this, prior DF outbreaks reported from Dire Dawa City were associated with DENV-1 and DEV-2 strains [7,20–23], although previous attempts to determine serotypes were restricted to a small number of DENV strains. Thus, the recurrent occurrence of dengue outbreaks associated with different DENV serotypes highlights the need for close monitoring and management of future DF outbreaks to effectively control and manage secondary infections.

The second possibility is that DENV-3III is more pathogenic, as suggested by prior research that connected outbreaks of DENV-3III to Dengue Hemorrhagic Fever (DHF) and Dengue Shock Syndrome (DSS) [24].

The third possibility is that the severe cases may be the result of co-infection with malaria parasites, as both infections are co-endemic in Dire Dawa City, and there is a chance that co-infection can worsen disease severity and outcomes. This is supported by the recent surge in malaria [25,26] and dengue cases [27,28] in Dire Dawa City, which posed significant challenges for healthcare providers. Given the co-endemicity and overlapping of symptoms (e.g., fever), there is a risk of co-infection with both dengue and malaria and misdiagnosing dengue as malaria (and vice versa). Thus, an integrated approach involving surveillance, education, and differential diagnosis is crucial for better patient care in Dire Dawa City Administration. Healthcare professionals need to be aware of the clinical features, diagnostic tests, and management protocols for malaria and dengue. In addition, the Dire Dawa City Administration should ensure that appropriate laboratory tests (such as rapid diagnostic tests and PCR) are available in all health facilities to enable accurate diagnosis of dengue and malaria and prompt treatment and management of severe cases.

Among the five distinct genotypes (I–V) of DENV-3, all 21 sequenced DENV strains from the present study belonged to DENV-3III, which is the most widespread and was associated with large outbreaks in Asia, Africa, and the Americas [29–31]. The first autochthonous case of DENV-3III in Africa was reported in Mozambique in 1985 [32]. Several man-made and natural factors, including (i) migration and conflict-related population displacement, (ii) frequent migration via trade routes that can spread the virus to new areas, and (iii) recurrent flooding that is made worse by climate change that provides ideal breeding grounds for Aedes mosquitoes, may have contributed to the rapid geographic expansion of DENV-3III in Africa, resulting in outbreaks in areas like Dire Dawa City [3,24,29,33,34].

Even though this work offers important insights into DENV strains responsible for the 2023 DF outbreak in Dire Dawa City where severe cases and deaths were recorded, it suffers limitations. First, the study's capacity to detect genetic variation was limited due to the use of a short genomic target (CprM, 511 base pairs) for the characterization of the outbreak strains. Expanding the genomic coverage could have enhanced our understanding of DENV diversity and relationships to other DENV-3III strains across the globe. Second, DENV CprM sequencing was performed on a small proportion of DF-suspected cases reported during the outbreak, and a larger sample size may have identified greater genetic diversity. Third, the study lacks information on other febrile illness-causing pathogens like malaria parasite examination with microscopy or rapid diagnostic test results for the studied DF-suspected cases, so cases of co-infection may have been missed. Future research addressing these limitations will therefore improve our comprehension of disease dynamics, provide a thorough understanding of DENV strains and their evolutionary history, expand our understanding of DENV outbreaks, and provide guidance for more effective preventative measures.

Historically, DENV infection and its severe manifestations have been underreported on the African continent [35]. However, this has changed recently. For instance, a study in Burkina Faso found that 33.5% of dengue cases were severe, with renal failure and severe bleeding being common complications [36]. Similarly, the present study reported a comparable proportion of dengue cases with warning signs that were predictors of severe dengue [37]. The severity and mortality of dengue infection in Africa may be associated with bacterial or other febrile coinfections (including malaria), comorbidities, or DENV serotypes [36,38]. However, the current study was hindered by the lack of crucial patient information, including immunosuppression, coinfections, and comorbidities, due to the unavailability of such data during the review of patient medical charts. Future studies should include comprehensive clinical and laboratory assessments of dengue patients to determine dengue fever severity.

5. Conclusions

In summary, our results demonstrate that the 2023 DF outbreak in Dire Dawa City, Ethiopia, was due to multiple co-circulating lineages of DENV-3III. The severe cases recorded during this outbreak may be due to either the genetic makeup of DENV-3III, secondary DENV infection, or co-infection of DENV and malaria. Additionally, the intercontinental transmission of DENV-3III underscores the importance of vigilant surveillance and preparedness to mitigate the impact of such outbreaks. Thus, collaborative efforts across borders are crucial for effective prevention and control strategies.

Supplementary Materials: The following supporting information can be downloaded at: https://www.mdpi.com/article/10.3390/v17030346/s1, Supplementary Figure S1. Maximum likelihood phylogenetic tree of DENV-3 genotype III (DENV-3III) strains isolated during the 2023 outbreak in Dire Dawa, Ethiopia with branch labels (the accession number, country of origin, and year of isolation). Supplementary Table S1: The 17 DENV strains for which high-quality CprM sequences were obtained and deposited into the GenBank database repository, along with their corresponding accession numbers, serotypes, genotypes, linages, and minor lineages, assigned to them by the newly developed lineage assignment tools.

Author Contributions: Conceptualization, A.A.N., T.G. and S.A.A.; study design, S.A.A., A.M. (Andargachew Mulu) and T.G.; methodology, A.A.N., A.M. (Andargachew Mulu), T.G., S.A.A. and D.H.A.; software, T.G. and D. D-G.; validation, T.G., D.R.-G. and A.P.; formal analysis, T.G., D.R.-G. and A.A. (Abaysew Ayele); laboratory investigations, D.H.A., T.K.W., G.B.T., A.H.B., D.T.G., E.A.D., E.Z.A., A.A.N., A.K.K., K.M. and R.J.; resources, A.M. (Andargachew Mulu), A.M. (Adane Mihret), S.A.A., A.A. (Alemseged Abdissa), G.T. and M.H.D.; field data and samples acquisition, E.Z.A., D.T.G. and A.K.K.; data curation, T.G., M.M.T. and A.A.N.; writing—original draft preparation, T.G. and A.A.N.; writing—review and editing, T.G., A.P., A.M. (Andargachew Mulu) and A.A.N.; visualization, T.G., D.R.-G., A.A. and A.P.; supervision, T.G., A.M. (Andargachew Mulu) and S.A.A. All authors have read and agreed to the published version of the manuscript.

Funding: This research received no external funding.

Institutional Review Board Statement: Ethical review and approval were waived for this study from the ALERT/AHRI Ethics Committee (PO-036-24) due to the fact it was conducted in response to the dengue fever outbreak.

Informed Consent Statement: Not applicable.

Data Availability Statement: The datasets generated and/or analyzed during the current study are available in the manuscript and the Supplementary Materials.

Acknowledgments: We would like to thank all study participants in this study. We are also grateful for the support and cooperation we received from physicians, nurses of Dire Dawa City health facilities, and administrative authorities of Dire Dawa City Administration Health Hureau and the health facilities administration. The authors acknowledge EPHI's support of the field study and AHRI and EPHI's reagent support of the molecular analysis.

Conflicts of Interest: The authors declare no conflicts of interest.

References

1. Dengue-Global Situation. Available online: https://www.who.int/emergencies/disease-outbreak-news/item/2023-DON498 (accessed on 7 April 2024).
2. Yang, X.; Quam, M.B.M.; Zhang, T.; Sang, S. Global burden for dengue and the evolving pattern in the past 30 years. *J. Travel Med.* **2021**, *28*, taab146. [CrossRef] [PubMed]
3. Geographical Expansion of Cases of Dengue and Chikungunya Beyond the Historical Areas of Transmission in the Region of the Americas. Available online: https://www.who.int/emergencies/disease-outbreak-news/item/2023-DON448 (accessed on 7 April 2024).
4. Roy, S.K.; Bhattacharjee, S. Dengue virus: Epidemiology, biology, and disease aetiology. *Can. J. Microbiol.* **2021**, *67*, 687–702. [CrossRef] [PubMed]
5. Dengue and Severe Dengue. Available online: https://www.who.int/news-room/fact-sheets/detail/dengue-and-severe-dengue (accessed on 25 February 2024).
6. Mekuriaw, W.; Kinde, S.; Kindu, B.; Mulualem, Y.; Hailu, G.; Gebresilassie, A.; Sisay, C.; Bekele, F.; Amare, H.; Wossen, M.; et al. Epidemiological, Entomological, and Climatological Investigation of the 2019 Dengue Fever Outbreak in Gewane District, Afar Region, North-East Ethiopia. *Insects* **2022**, *13*, 1066. [CrossRef] [PubMed]
7. Gutu, M.A.; Bekele, A.; Seid, Y.; Mohammed, Y.; Gemechu, F.; Woyessa, A.B.; Tayachew, A.; Dugasa, Y.; Gizachew, L.; Idosa, M.; et al. Another dengue fever outbreak in Eastern Ethiopia—An emerging public health threat. *PLoS Neglected Trop. Dis.* **2021**, *15*, e0008992. [CrossRef]
8. Mesfin, Z.; Ali, A.; Abagero, A.; Asefa, Z. Dengue Fever Outbreak Investigation in Werder Town, Dollo Zone, Somali Region, Ethiopia. *Infect. Drug Resist.* **2022**, *15*, 7207–7217. [CrossRef]
9. Mohan, A.; Fakhor, H.; Nimavat, N.; Wara, U.U.; Lal, P.M.; Costa, A.C.d.S.; Ahmad, S.; Essar, M.Y. Dengue and COVID-19: A risk of coepidemic in Ethiopia. *J. Med. Virol.* **2021**, *93*, 5680–5681. [CrossRef]
10. Sisay, C.; Tadesse, A.; Workie, F.; Yizengaw, A.; Ali, A.; Yusuf, J.; Abose, E.; Tsega, D.; Kinde, S.; Tadesse, H. Description of Serotype 3 Dengue Fever Virus: Clinical, Surveillance and Geographical Expansion in the Northeastern Ethiopia. *Sci. J. Public Health* **2023**, *11*, 123–131. [CrossRef]
11. Htun, T.P.; Xiong, Z.; Pang, J. Clinical signs and symptoms associated with WHO severe dengue classification: A systematic review and meta-analysis. *Emerg. Microbes Infect.* **2021**, *10*, 1116–1128. [CrossRef]
12. Lanciotti, R.S.; Calisher, C.H.; Gubler, D.J.; Chang, G.J.; Vorndam, A.V. Rapid detection and typing of dengue viruses from clinical samples by using reverse transcriptase-polymerase chain reaction. *J. Clin. Microbiol.* **1992**, *30*, 545–551. [CrossRef]
13. Hill, V.; Cleemput, S.; Pereira, J.S.; Gifford, R.J.; Fonseca, V.; Tegally, H.; Brito, A.F.; Ribeiro, G.; de Souza, V.C.; Brcko, I.C.; et al. A New Lineage Nomenclature to Aid Genomic Surveillance of Dengue Virus. *PLoS Biol.* **2024**, *22*, e3002834. [CrossRef]
14. Katoh, K.; Standley, D.M. MAFFT Multiple Sequence Alignment Software Version 7: Improvements in Performance and Usability. *Mol. Biol. Evol.* **2013**, *30*, 772–780. [CrossRef] [PubMed]
15. Nguyen, L.-T.; Schmidt, H.A.; von Haeseler, A.; Minh, B.Q. IQ-TREE: A Fast and Effective Stochastic Algorithm for Estimating Maximum-Likelihood Phylogenies. *Mol. Biol. Evol.* **2015**, *32*, 268–274. [CrossRef] [PubMed]
16. Letunic, I.; Bork, P. Interactive Tree Of Life (iTOL) v5: An online tool for phylogenetic tree display and annotation. *Nucleic Acids Res.* **2021**, *49*, W293–W296. [CrossRef] [PubMed]
17. Mekonnen, F.; Khan, B.A.; Nibret, E.; Munshea, A.; Tsega, D.; Endalamaw, D.; Tadesse, S.; Yismaw, G.; Lankir, D.; Ali, J.; et al. Introduction of Dengue Virus Serotype 3 in the Afar Region, Ethiopia. *Emerg. Microbes Infect.* **2024**, *13*, 2429653. [CrossRef] [PubMed]
18. Eldigail, M.H.; Abubaker, H.A.; Khalid, F.A.; Abdallah, T.M.; Musa, H.H.; Ahmed, M.E.; Adam, G.K.; Elbashir, M.I.; Aradaib, I.E. Association of genotype III of dengue virus serotype 3 with disease outbreak in Eastern Sudan, 2019. *Virol. J.* **2020**, *17*, 118. [CrossRef]
19. Muthanje, E.M.; Kimita, G.; Nyataya, J.; Njue, W.; Mulili, C.; Mugweru, J.; Mutai, B.; Kituyi, S.N.; Waitumbi, J. March 2019 dengue fever outbreak at the Kenyan south coast involving dengue virus serotype 3, genotypes III and V. *PLoS Glob. Public Health* **2022**, *2*, e0000122. [CrossRef]
20. Biru, M.; Geleta, D.; Tesfaye, N.; Aleyu, M.; Tayachew, A. Dengue Fever Outbreak Investigation and Response in Dire Dawa City Administration, Ethiopia, 2017. *J. Med.* **2020**, *63*, 23–27. [CrossRef]
21. Zerfu, B.; Kassa, T.; Legesse, M. Epidemiology, biology, pathogenesis, clinical manifestations, and diagnosis of dengue virus infection, and its trend in Ethiopia: A comprehensive literature review. *Trop. Med. Health* **2023**, *51*, 11. [CrossRef]
22. Sisay, C.; Waldetensai, A.; Seyoum, M.; Asrat, Y.; Tayachew, A.; Wossen, M.; Keneni, D.; Daniel, T.; Esmael, N.; Abdulhamid, I. Detection of serotype 1-Dengue fever outbreak in Dire Dawa city, Eastern Ethiopia. *Ethiop. J. Public Health Nutr. (EJPHN)* **2022**, *5*, 49–54.

23. Tsegaye, M.M.; Mekonnen, A.T.; Gebretsion, D.T.; Gelanew, T.; Alemayehu, D.H.; Tefera, D.A.; Woldemichael, T.S.; Getaneh, B.A. Abera, E.K.; Jebessa, G.G.; et al. Predominance of Dengue Virus Serotype-1/Genotype-I in Eastern and Southeastern Ethiopia *Viruses* **2024**, *16*, 1334. [CrossRef]
24. Messer, W.B.; Gubler, D.J.; Harris, E.; Sivananthan, K.; Silva AM, de. Emergence and Global Spread of a Dengue Serotype 3 Subtype III Virus. *Emerg. Infect. Dis.* **2003**, *9*, 800–809. [CrossRef] [PubMed]
25. A Malaria Outbreak in Ethiopia Came from an Invasive Asian Mosquito. 2022. Available online: https://www.sciencenews.org/article/malaria-outbreak-ethiopia-invasive-mosquito-asia (accessed on 11 March 2024).
26. Emiru, T.; Getachew, D.; Murphy, M.; Sedda, L.; Ejigu, L.A.; Bulto, M.G.; Byrne, I.; Demisse, M.; Abdo, M.; Chali, W.; et al. Evidence for a role of Anopheles stephensi in the spread of drug- and diagnosis-resistant malaria in Africa. *Nat. Med.* **2023**, *29*, 3203–3211. [CrossRef]
27. Degife, L.H.; Worku, Y.; Belay, D.; Bekele, A.; Hailemariam, Z. Factors associated with dengue fever outbreak in Dire Dawa administration city, October, 2015, Ethiopia—Case control study. *BMC Public Health* **2019**, *19*, 650. [CrossRef] [PubMed]
28. Nigussie, E.; Atlaw, D.; Negash, G.; Gezahegn, H.; Baressa, G.; Tasew, A.; Zembaba, D. A dengue virus infection in Ethiopia: A systematic review and meta-analysis. *BMC Infect. Dis.* **2024**, *24*, 297. [CrossRef]
29. Maquart, M. Circulation of Dengue Virus Type 3 Genotype III in Africa Since 2008. *J. Hum. Virol. Retrovirol.* **2016**, *4*, 1–11. [CrossRef]
30. Waman, V.P.; Kale, M.M.; Kulkarni-Kale, U. Genetic diversity and evolution of dengue virus serotype 3: A comparative genomics study. *Infect. Genet. Evol.* **2017**, *49*, 234–240. [CrossRef]
31. Titir, S.R.; Paul, S.K.; Ahmed, S.; Haque, N.; Nasreen, S.A.; Hossain, K.S.; Ahmad, F.U.; Nila, S.S.; Khanam, J.; Nowsher, N.; et al. Nationwide Distribution of Dengue Virus Type 3 (DENV-3) Genotype I and Emergence of DENV-3 Genotype III During the 2019 Outbreak in Bangladesh. *Trop. Med. Infect. Dis.* **2021**, *6*, 58. [CrossRef]
32. Gubler, D.J.; Sather, G.E.; Kuno, G.; Cabral, J.R. Dengue 3 Virus Transmission in Africa. *Am. J. Trop. Med. Hyg.* **1986**, *35*, 1280–1284. [CrossRef]
33. Amarasinghe, A.; Kuritsky, J.N.; Letson, G.W.; Margolis, H.S. Dengue Virus Infection in Africa. *Emerg. Infect. Dis.* **2011**, *17*, 1349–1354. [CrossRef]
34. Eltom, K.; Enan, K.; El Hussein, A.R.M.; Elkhidir, I.M. Dengue Virus Infection in Sub-Saharan Africa Between 2010 and 2020: A Systematic Review and Meta-Analysis. *Front. Cell Infect. Microbiol.* **2021**, *11*, 678945. [CrossRef]
35. Gainor, E.M.; Harris, E.; LaBeaud, A.D. Uncovering the Burden of Dengue in Africa: Considerations on Magnitude, Misdiagnosis and Ancestry. *Viruses* **2022**, *14*, 233. [CrossRef] [PubMed]
36. Sondo, A.K.; Diendéré, E.A.; Meda, B.I.; Diallo, I.; Zoungrana, J.; Poda, A.; Manga, N.M.; Bicaba, B.; Gnamou, A.; Kagoné, C.J.; et al. Severe Dengue in Adults and Children, Ouagadougou (Burkina Faso), West Africa, October 2015–January 2017. *IJID Re*[illegible] **2021**, *1*, 53–59. [CrossRef] [PubMed]
37. Tsheten, T.; Clements, A.C.A.; Gray, D.J.; Adhikary, R.K.; Furuya-Kanamori, L.; Wangdi, K. Clinical Predictors of Severe Dengue: A Systematic Review and Meta-Analysis. *Infect. Dis. Poverty* **2021**, *10*, 123. [CrossRef]
38. Gebremariam, T.T.; Schallig, H.D.F.H.; Kurmane, Z.M.; Danquah, J.B. Increasing Prevalence of Malaria and Acute Dengue Virus Coinfection in Africa: A Meta-Analysis and Meta-Regression of Cross-Sectional Studies. *Malar. J.* **2023**, *22*, 300. [CrossRef]

viruses

MDPI

Communication

Factors Associated with Chronic Chikungunya in Vitória, Espírito Santo State, Brazil, Between 2016 and 2020

Creuza Rachel Vicente [1,2,*], Luana Santos Louro [3], Nicolli Ribeiro de Jesus [2], Danielle Torres dos Santos Lopes [2], Aline Souza Areias Cabidelle [4], Crispim Cerutti Junior [1,2], Angelica Espinosa Barbosa Miranda [1,2], Iuri Drumond Louro [5,6], Debora Dummer Meira [5,6] and Kuan Rong Chan [7,*]

1 Department of Social Medicine, Federal University of Espírito Santo, Vitória 29047-105, Espírito Santo State, Brazil; fil.cris@terra.com.br (C.C.J.); angelica.miranda@ufes.br (A.E.B.M.)
2 Post-Graduate Program in Infectious Diseases, Federal University of Espírito Santo, Vitória 29047-105, Espírito Santo State, Brazil; nicollirj@gmail.com (N.R.d.J.); danitsl@yahoo.com.br (D.T.d.S.L.)
3 Medical School, Federal University of Espírito Santo, Vitória 29047-105, Espírito Santo State, Brazil; luana.louro@edu.ufes.br
4 Health Surveillance Service, Department of Health, Vitória 29017-010, Espírito Santo State, Brazil; areiasaline927@gmail.com
5 Department of Animal Biology, Federal University of Espírito Santo, Vitória 29075-910, Espírito Santo State, Brazil; iuri.louro@ufes.br (I.D.L.); debora.dummer.meira@gmail.com (D.D.M.)
6 Post-Graduate Program in Biotechnology, Federal University of Espírito Santo, Vitória 29047-105, Espírito Santo State, Brazil
7 Program in Emerging Infectious Diseases, Duke-NUS Medical School, Singapore 169857, Singapore
* Correspondence: vicentecrachel@gmail.com (C.R.V.); kuanrong.chan@duke-nus.edu.sg (K.R.C.)

Abstract: Chikungunya patients may develop chronic joint pain that can persist for months to years. This study aimed to determine the factors associated with Chikungunya chronicity. This case–control study involved data from patients with laboratory-confirmed Chikungunya reported from March 2016 to December 2020 in Vitória, Espírito Santo state, Brazil. The data were accessed from compulsory notification databases (SINAN and eSUS VS) and electronic medical reports (Rede Bem-Estar). For each patient who developed chronic symptoms, we included a control patient who did not develop chronic symptoms by random sampling. A total of 183 chronic and 183 non-chronic patients were included in the study. Most of them were female (73.2%), with a median age of 49.5 years (interquartile range = 37–61), and presented fever (89.6%), myalgia (89.6%), arthralgia (89.3%), and headache (82.0%). Chronic patients were older (median = 53; interquartile range = 41–61) than non-chronic cases (median = 46; interquartile age = 31–61) (OR = 0.979, 95% CI = 0.968–0.991) and more frequently presented nausea (58.5% vs. 40.4%; OR = 2.109, 95% CI = 1.374–3.238), and leukopenia (20.2% vs. 10.9%; OR = 2.060, 95% CI = 1.122–3.779). Therefore, these characteristics should be monitored for the better clinical management of cases prone to chronicity.

Keywords: chikungunya fever; chronic pain; arthralgia; clinical epidemiology

1. Introduction

Chikungunya is endemic in about 114 African, Asian, and American countries and territories, with sporadic outbreaks in other locations and an estimated number of annual symptomatic cases ranging between 52,774 and 328,943 [1]. In the Americas, the disease emerged in 2013, and in 10 years, the region had the highest global burden, with more than 3.7 million suspected cases reported [1]. Since 2016, Brazil has been the epicenter of Chikungunya epidemics in the Americas, with the highest number of reports, accounting for more than 1.5 million cases from 2016 to 2023 [2].

Chikungunya is a systemic infection caused by the Chikungunya virus (CHIKV) (Alphavirus genus, Togaviridae family), which is transmitted by mosquitoes of the Aedes genus, especially the species *Aedes aegypti*. There are four genotypes of CHIKV, namely West African, East/Central/South African (ECSA), Asian, and Indian Ocean [3].

The most typical manifestation of Chikungunya is intense and debilitating joint pain, usually affecting different sites, impacting daily activities, and resulting in mental distress [4]. In the acute phase of the disease, the patient can present with pain in the back, muscles, and joints; headache; joint swelling; rash; and fatigue. The chronic phase occurs when joint manifestations persist for three or more months, having periods of remission and recurrence, with 42.5% of Chikungunya cases developing arthritis or arthralgia, 25% inflammatory rheumatism, and some reporting musculoskeletal stiffness, with frequencies varying in different populations [1,4–8]. The frequencies of chronic symptoms may differ according to CHIKV lineages, with higher prevalence reported in the ECSA genotype [9].

The pathogenesis of chronic Chikungunya has not yet been fully clarified. Host characteristics could contribute to chronicity, such as female sex, aging, diabetes, hypertension, and severe pain in the acute stage [10]. The absence of specific drugs for treatment and licensed vaccines poses additional challenges for the disease's clinical management and prevention [11].

In Brazil, the high prevalence of chronic arthralgia in Chikungunya infections demands long-term multidisciplinary medical care, burdening health systems and leading to social impacts [12]. Therefore, this study aims to identify the potential factors associated with chronic Chikungunya to improve the identification and clinical management of cases prone to chronicity.

2. Materials and Methods

2.1. Study Design

This case–control study involves data on Chikungunya reports in residents of Vitória municipality, Espírito Santo state, Brazil, with clinical manifestations beginning between 1 March 2016 and 31 December 2020. The outcome assessed was the progression to the chronic phase of Chikungunya. A total of 183 chronic cases included in the study were matched with 183 controls who did not develop chronic symptoms. The controls were selected by simple random sampling.

2.2. Study Area and Population

Vitória is Espírito Santo's capital on southeast Brazil's coast. It has 97.123 km^2 of insular and continental areas and a tropical humid climate. The population in Vitória in 2022 was 322,869 people, with a population density of 3324.33 inhabitants per km^2 [13].

2.3. Data Source

Data on Chikungunya compulsory reports from 1 March 2016 to 31 March 2020 were collected from the Notifiable Diseases Information System (SINAN) and, from 1 April 2020 to 31 December 2020, from the eSUS Health Surveillance (eSUS-VS). Data extraction occurred on 17 September 2021. In addition, the medical records from Rede Bem-Estar were accessed to follow patients' clinical outcomes.

2.4. Inclusion and Exclusion Criteria

The study included laboratory-confirmed cases of Chikungunya in viral isolation, RT-PCR or ELISA IgM, who resided in Vitória. Reports without data on the Chikungunya outcome and duplicates were excluded.

2.5. Variables

The independent variables comprised age, sex, Chikungunya clinical manifestations, pre-existing diseases, and hospitalization. The outcome was chronic Chikungunya, characterized by the persistence of clinical manifestations for over three months. Clinical manifestations in the acute phase included fever, intense polyarthralgia, back and head pain, rash, and fatigue [14].

2.6. Sample Size

Considering a 95% confidence interval, a test power of 80%, an exposure in 75% of the controls, a proportion of one control for each case, and a three-fold difference in exposure between the groups, the minimum sample size was 113 cases and 113 controls. The calculation was performed using Statcalc (EpiInfoTM software version 7.2.2.6, Centers for Disease Control and Prevention, Atlanta, GA, USA).

2.7. Statistical Analysis

Categorical variables were described by absolute and relative frequencies and compared between the groups using Pearson's chi-square test or Fisher's exact test. Continuous variables were described as means and interquartile ranges and compared between the groups using the Mann–Whitney test. The variables with statistically significant differences were analyzed by binary logistic regression using the Forward Stepwise method. The results were presented as Odds Ratios (ORs) with a 95% confidence interval (95% CI), and a p-value ≤ 0.05 was considered significant. The analyses were performed using SPSS version 20.

2.8. Ethical Considerations

The Research Ethics Committee of the Federal University of Espírito Santo approved the study protocol under opinion number 4,393,656.

3. Results

Between March 2016 and December 2020, 3216 cases of Chikungunya with laboratory confirmation were identified in Vitória. Of these, 19 were excluded from the study—15 due to a lack of outcome data and 4 because they were duplicates. Thus, 3197 patients were eligible, of which, 3176 were confirmed by IgM serology in the first sample, 15 by IgM serology in the second sample, 15 by RT-PCR, and 1 by viral isolation. Some patients underwent more than one laboratory confirmation test. The data completeness of the variables in the included cases varied from 99.7% to 100%.

Of the eligible cases, 183 (5.7%) were classified as chronic; therefore, 183 non-chronic patients were randomly selected from the 3,014 who presented this classification. Most of the study population was female (73.2%), with ages ranging from 0 to 89 years and a median of 49.5 years (interquartile range = 37–61). The most frequent pre-existing diseases were arterial hypertension (31.1%) and diabetes (15.6%). Fever (89.6%), myalgia (89.6%), arthralgia (89.3%), and headache (82.0%) were the main Chikungunya clinical manifestations, and 2.6% of patients required hospitalization (Table 1).

Patients with chronic Chikungunya were significantly older than those who had non-chronic Chikungunya (p-value = 0.001); reported more hypertension (p-value = 0.042); and presented more nausea (p-value = 0.001), vomiting (p-value = 0.014), leukopenia (p-value = 0.014), and back pain (p-value = 0.057). Hospitalization (p-value = 0.095) was not associated with chronicity (Table 1).

In the multivariate analysis, nausea and leukopenia were associated with chronic presentation. On the other hand, there was a small protective effect for chronicity for age, which was lower among non-chronic patients (Table 2).

Binary logistic regression was carried out using the Forward Stepwise method. Variables not included in the model were vomiting, arterial hypertension, and back pain. X^2 (3) = 28.588, p-value = 0.000, R^2 Nagelkerke = 0.100, * Odds Ratio adjusted with 95% confidence interval. Sensitivity = 60.1%, specificity = 61.2%, accuracy = 60.7%, precision = 60.8%. F1-score = 60.4%, AUC = 0.6865.

Table 1. Characteristics of Chikungunya cases according to chronicity outcome.

Characteristic	Chronic n = 183 n (%)	Non-Chronic n = 183 n (%)	Total n = 366 n (%)	*p*-Value
Sex				
Female	140 (76.5%)	128 (69.9%)	268 (73.2%)	0.157 ł
Male	43 (23.5%)	55 (30.1%)	98 (26.8%)	
Age				
Median (interquartile range)	53 (41–61)	46 (31–61)	49.5 (37–61)	0.006 #
Pre-existing diseases				
Diabetes	33/182 (18.1%)	24 (13.1%)	57 (15.6%)	0.187 ł
Liver disease	5/182 (2.7%)	1 (0.5%)	6 (1.6%)	0.121 ¶
Arterial hypertension	66 (36.1%)	48 (26.2%)	114 (31.1%)	0.042 ł
Acid-peptic disease	6 (3.3%)	3 (1.6%)	9 (2.5%)	0.502 ¶
Clinical manifestation				
Fever	163 (89.1%)	165 (90.2%)	328 (89.6%)	0.732 ł
Myalgia	162 (88.5%)	166 (90.7%)	328 (89.6%)	0.493 ł
Headache	154 (84.2%)	146 (79.8%)	300 (82.0%)	0.277 ł
Rash	66 (36.1%)	68 (37.2%)	134 (36.6%)	0.828 ł
Vomiting	42 (23.0%)	24 (13.1%)	66 (18.0%)	0.014 ł
Nausea	107 (58.5%)	74 (40.4%)	181 (49.5%)	0.001 ł
Back pain	133 (72.7%)	116 (63.4%)	249 (68.0%)	0.057 ł
Conjunctivitis	19 (10.4%)	22 (12.0%)	41 (11.2%)	0.619 ł
Arthritis	90 (49.2%)	86 (47.0%)	176 (48.1%)	0.676 ł
Arthralgia	166 (90.7%)	161 (88.0%)	327 (89.3%)	0.397 ł
Petechiae	26 (14.2%)	30 (16.4%)	56 (15.3%)	0.561 ł
Leukopenia	37 (20.2%)	20 (10.9%)	57 (15.6%)	0.014 ł
Positive tourniquet test	26 (14.2%)	28 (15.3%)	54 (14.8%)	0.768 ł
Retroorbital pain	82 (44.8%)	72 (39.3%)	154 (42.1%)	0.290 ł
Hospitalization	7 (4.2%)	2 (1.1%)	9 (2.6%)	0.095 ¶

ł Pearson's Chi-square test, ¶ Fisher's exact test, # Mann–Whitney test.

Table 2. Characteristics associated with the chronic Chikungunya.

Characteristic	OR *	95% Confidence Interval Lower Limit	Upper Limit
Age	0.979	0.968	0.991
Nausea	2.109	1.374	3.238
Leukopenia	2.060	1.122	3.779

4. Discussion

This study described the profile of those affected by Chikungunya, particularly regarding the evolution to chronicity and its associated factors. The demographic characteristics of those affected were like those in other settings in Brazil, with more reports of the disease in adults and women, representing the medical care-seeking behavior, not only the incidence [15]. Nevertheless, the influence of sex on immune response could have contributed to a higher proportion of Chikungunya-symptomatic females, with hormones affecting the production of proinflammatory cytokines and antibodies [16]. Aging was associated with chronicity, as previously reported, especially in patients over 40 years old [6,16]. Thus, age is a relevant factor for disease progression.

Arterial hypertension and diabetes were the most common pre-existing chronic diseases in Chikungunya cases since they are also the most frequent noncommunicable diseases in the Brazilian population. Both comorbidities were previously reported to affect 20% to 30% of patients with CHIKV infection [17]. In addition, hypertension was significantly more frequent in chronic Chikungunya, although not considerably significantly associated in the final model. Age may also confound this finding since hypertension is related to aging [18].

The main acute clinical manifestations of Chikungunya were fever, myalgia, arthralgia, and headache, comparable in frequency to those found in a systematic review that showed 92% of cases presenting swelling and joint pain, 85% fever, 52% myalgia, and 30% to 40% neurological symptoms, such as headache [19]. The CHIKV tropism to joints, muscles, bones, and nerves, accompanied by the release of inflammatory and pyrogenic cytokines, such as IL-6, IL1b, and TNFα, likely contributed to the emergence of these signs and symptoms [20]. Notably, we detected that chronic patients presented significantly more symptoms, including vomiting and back pain, and especially nausea and leukopenia, which were the most significantly associated factors for chronicity in the final model. These findings are consistent with previous studies demonstrating that severe pain in the acute phase of the disease predicts its chronicity [6,16]. Moreover, vomiting was associated with an increased risk of severity and death by CHIKV infection [21]. Leukopenia was also previously identified as a risk factor for hospitalization, severity, and deaths in patients with Chikungunya, although no previous studies linked it with chronic cases [21,22]. While it remains unknown how increased nausea and leukopenia during the acute phase are linked to the increased occurrence of chronic symptoms, one possible explanation may be related to the increased use of opioids to manage exacerbated pain that increases the side effects and promotes CHIKV disease pathogenesis.

Hospitalization was infrequent and not associated with chronic Chikungunya. In Vitória, hospitalization was previously related to specific age groups, such as those lower than two and over 65 years, and in patients with complications such as thrombocytopenia, diarrhea, and syncope [23]. Concerning the epidemiology of CHIKV in our municipality, the ECSA genotype was identified in 2016 and 2017. In other areas of Brazil, the Asian genotype was also identified [3,24]. Although we did not evaluate the viral genotypes that were circulating in Vitória in this study, the prevalence of chronic cases in ECSA is expected to vary from 40% to 60%, indicating a possibility of under-reporting of the condition in the municipality population and the necessity of improved follow-up of the patients and updating the report forms [9].

The study limitations are mainly related to secondary data from the compulsory reports and the possibility of information bias. Nevertheless, medical records were consulted to improve data quality and confirm the outcome. However, the health services did not conduct follow-up, and only patients who spontaneously returned to healthcare were reported as chronic. Therefore, we cannot discard the misclassification of patients considered non-chronic. The strict inclusion of laboratory-confirmed cases prevented the inclusion of other conditions with similar clinical manifestations, such as dengue [25,26]. In addition, we did not access other possible determinants of chronicity, such as viral and immunological factors. Cohort studies are suggested to confirm the presented findings, considering the follow-up of leukocytes and therapeutic approaches, such as the use of opioids. In addition, future prospective studies should also evaluate viral and immunological factors in Chikungunya chronicity, including how sex contributes to clinical manifestations.

This study reinforces the necessity of monitoring Chikungunya patients, especially those older, in need of opioid use due to severe pain or with laboratory alterations, such as leukopenia, for the better identification and clinical management of cases prone to chronicity.

5. Conclusions

Chronic Chikungunya was associated with older age, the presence of nausea, and leukopenia, suggesting the need for the longer-term monitoring of patients with these characteristics to improve diagnosis and treatment.

Author Contributions: Conceptualization, C.R.V. and A.S.A.C.; Methodology, C.R.V.; Validation, K.R.C.; Formal Analysis, C.R.V.; Investigation, C.R.V.; Resources, C.R.V. and A.S.A.C.; Data Curation, L.S.L., N.R.d.J., D.T.d.S.L. and A.S.A.C.; Writing—Original Draft Preparation, C.R.V.; Writing—Review and Editing, L.S.L., N.R.d.J., D.T.d.S.L., A.S.A.C., C.C.J., A.E.B.M., I.D.L., D.D.M. and K.R.C.; Visualization, C.R.V.; Project Administration, C.R.V. All authors have read and agreed to the published version of the manuscript.

Funding: This research was funded by the "Brazilian National Council for Scientific and Technological Development" (CNPq) [grant number 442106/2023-8 to CRV], the "Brazilian Coordination for the Improvement of Higher Education Personnel" (CAPES) [grant number 88887.927805/2023-00 to NRJ]. KRC is funded by the National Medical Research Council [grant number MOH-000610-00 to KRC].

Institutional Review Board Statement: The study was conducted according to the guidelines of the Declaration of Helsinki and approved by the Research Ethics Committee of the Federal University of Espírito Santo (protocol code 4393656/11 November 2020).

Informed Consent Statement: Patient consent was waived due to the use of anonymized databanks.

Data Availability Statement: The data presented in this study are available on request from the corresponding author due to the presence of patient data.

Acknowledgments: The authors thank Gustavo Macêdo de Carvalho for the suggestions on the data analysis.

Conflicts of Interest: The authors declare no conflicts of interest.

References

1. Puntasecca, C.J.; King, C.H.; LaBeaud, A.D. Measuring the global burden of Chikungunya and Zika viruses: A systematic review. *PLoS Negl. Trop. Dis.* **2021**, *15*, e0009055. [CrossRef] [PubMed]
2. de Souza, W.M.; Ribeiro, G.S.; de Lima, S.T.S.; de Jesus, R.; Moreira, F.R.R.; Whittaker, C.; Sallum, M.A.M.; Carrington, C.V.F.; Sabino, E.C.; Kitron, U.; et al. Chikungunya: A decade of burden in the Americas. *Lancet Reg. Health Am.* **2024**, *30*, 100673 [CrossRef] [PubMed]
3. Giovanetti, M.; Vazquez, C.; Lima, M.; Castro, E.; Rojas, A.; Gomez de la Fuente, A.; Aquino, C.; Cantero, C.; Fleitas, F.; Torales, J.; et al. Rapid epidemic expansion of Chikungunya virus East/Central/South African lineage, Paraguay. *Emerg. Infect. Dis.* **2023**, *29*, 1859–1863. [CrossRef] [PubMed]
4. Watson, H.; Tritsch, S.R.; Encinales, L.; Cadena, A.; Cure, C.; Ramirez, A.P.; Mendoza, A.R.; Chang, A.Y. Stiffness, pain, and joint counts in chronic Chikungunya disease: Relevance to disability and quality of life. *Clin. Rheumatol.* **2020**, *39*, 1679–1686 [CrossRef]
5. Rodríguez-Morales, A.J.; Cardona-Ospina, J.A.; Urbano-Garzón, S.F.; Hurtado-Zapata, J.S. Prevalence of post-Chikungunya infection chronic inflammatory arthritis: A systematic review and meta-analysis. *Arth. Care Res.* **2016**, *68*, 1849–1858. [CrossRef]
6. Cunha, R.V.; Trinta, K.S. Chikungunya virus: Clinical aspects and treatment—A review. *Mem. Inst. Oswaldo Cruz* **2017**, *112* 523–531. [CrossRef]
7. Sosa-Martínez, M.J.; Orea-Flores, M.; Vázquez-Cruz, I.; Palacios-Castillo, V.; Juanico-Morales, G.; Pérez-Mijangos, L. Caracterización de las manifestaciones clínicas crónicas en pacientes con fiebre Chikunguña [Characterization of chronic clinical manifestations in patients with Chikungunya fever]. *Rev. Med. Inst. Mex. Seguro Soc.* **2018**, *56*, 239–245.
8. Tritsch, S.R.; Encinales, L.; Pacheco, N.; Cadena, A.; Cure, C.; McMahon, E.; Watson, H.; Porras Ramirez, A.; Mendoza, A.R.; Li G.; et al. Chronic joint pain 3 years after Chikungunya virus infection largely characterized by relapsing-remitting symptoms. *J. Rheumatol.* **2020**, *47*, 1267–1274. [CrossRef]
9. Paixão, E.S.; Rodrigues, L.C.; Costa, M.D.C.N.; Itaparica, M.; Barreto, F.; Gérardin, P.; Teixeira, M.G. Chikungunya chronic disease: A systematic review and meta-analysis. *Trans. R. Soc. Trop. Med. Hyg.* **2018**, *112*, 301–316. [CrossRef]
10. Noor, F.M.; Hossain, M.B.; Islam, Q.T. Prevalence of and risk factors for long-term disabilities following Chikungunya virus disease: A meta-analysis. *Travel Med. Infect. Dis.* **2020**, *35*, 101618. [CrossRef]
11. de Souza, W.M.; de Lima, S.T.S.; Simões Mello, L.M.; Candido, D.S.; Buss, L.; Whittaker, C.; Claro, I.M.; Chandradeva, N.; Granja F.; de Jesus, R.; et al. Spatiotemporal dynamics and recurrence of Chikungunya virus in Brazil: An epidemiological study. *Lancet Microbe* **2023**, *4*, e319–e329. [CrossRef] [PubMed]
12. Silva, M.M.O.; Kikuti, M.; Anjos, R.O.; Portilho, M.M.; Santos, V.C.; Gonçalves, T.S.F.; Tauro, L.B.; Moreira, P.S.S.; Jacob Nascimento, L.C.; Santana, P.M.; et al. Risk of chronic arthralgia and impact of pain on daily activities in a cohort of patients with Chikungunya virus infection from Brazil. *Int. J. Infect. Dis.* **2021**, *105*, 608–616. [CrossRef] [PubMed]
13. Instituto Brasileiro de Geografia e Estatística. Cidades e Estados—Vitória. Available online: https://www.ibge.gov.br/cidades-e-estados/es/vitoria.html (accessed on 15 April 2024).
14. Brasil, Ministério da Saúde. *Chikungunya: Manejo Clínico*, 1st ed.; Ministério da Saúde: Brasília, Brazil, 2017.

15. Silva, A.D.C.; Silva, A.D.C.; de Castro, P.A.S.V.; Ávila, I.R.; Bezerra, J.M.T. Prevalence and epidemiological aspects of Chikungunya fever in states of the Northeast region of Brazil: A systematic review. *Acta Trop.* **2023**, *241*, 106872. [CrossRef] [PubMed]
16. Bertolotti, A.; Thioune, M.; Abel, S.; Belrose, G.; Calmont, I.; Césaire, R.; Cervantes, M.; Fagour, L.; Javelle, É.; Lebris, C.; et al. Prevalence of chronic Chikungunya and associated risks factors in the French West Indies (La Martinique): A prospective cohort study. *PLoS Negl. Trop. Dis.* **2020**, *14*, e0007327. [CrossRef]
17. Badawi, A.; Ryoo, S.G.; Vasileva, D.; Yaghoubi, S. Prevalence of chronic comorbidities in Chikungunya: A systematic review and meta-analysis. *Int. J. Infect. Dis.* **2018**, *67*, 107–113. [CrossRef]
18. Vicente, C.R.; Cerutti Junior, C.; Fröschl, G.; Romano, C.M.; Cabidelle, A.S.; Herbinger, K.H. Influence of demographics on clinical outcome of Dengue: A cross-sectional study of 6703 confirmed cases in Vitória, Espírito Santo State, Brazil. *Epidemiol. Infect.* **2017**, *145*, 46–53. [CrossRef]
19. Bartholomeeusen, K.; Daniel, M.; LaBeaud, D.A.; Gasque, P.; Peeling, R.W.; Stephenson, K.E.; Ng, L.F.P.; Ariën, K.K. Chikungunya fever. *Nat. Rev. Dis. Prim.* **2023**, *9*, 17. [CrossRef]
20. Matusali, G.; Colavita, F.; Bordi, L.; Lalle, E.; Ippolito, G.; Capobianchi, M.R.; Castilletti, C. Tropism of the Chikungunya virus. *Viruses* **2019**, *11*, 175. [CrossRef]
21. Silva Junior, G.B.D.; Pinto, J.R.; Mota, R.M.S.; Pires Neto, R.D.J.; Daher, E.F. Risk factors for death among patients with Chikungunya virus infection during the outbreak in northeast Brazil, 2016–2017. *Trans. R. Soc. Trop. Med. Hyg.* **2019**, *113*, 221–226. [CrossRef]
22. Pinto, J.R.; Silva Junior, G.B.D.; Mota, R.M.S.; Martins, P.; Santos, A.K.T.; Moura, D.C.N.; Pires Neto, R.D.J.; Daher, E.F. Clinical profile and factors associated with hospitalization during a Chikungunya epidemic in Ceará, Brazil. *Rev. Soc. Bras. Med. Trop.* **2019**, *52*, e20190167. [CrossRef]
23. Lopes, D.T.S.; Cerutti Junior, C.; Areias Cabidelle, A.; Espinosa Miranda, A.; Drumond Louro, I.; Pamplona de Góes Cavalcanti, L.; Vicente, C.R. Factors associated with hospitalization in the acute phase of Chikungunya. *PLoS ONE* **2023**, *18*, e0296131.
24. Ventorim, D.P. Diversidade Genética de Chikungunya no Estado do Espírito Santo. Master's Thesis, Universidade Federal do Espírito Santo, Vitória, Brazil, 5 March 2018.
25. Bonifay, T.; Lienne, J.F.; Bagoée, C.; Santa, F.; Vesin, G.; Walter, G.; Nacher, M.; Vaserman, N.; Djossou, F.; Epelboin, L. Prevalence and risk factors of post Chikungunya rheumatic musculoskeletal disorders: A prospective follow-up study in French Guiana. *Eur. J. Clin. Microbiol. Infect. Dis.* **2018**, *37*, 2159–2164. [CrossRef] [PubMed]
26. Almeida, I.F.; Codeço, C.T.; Lana, R.M.; Bastos, L.S.; de Souza Oliveira, S.; da Cruz Ferreira, D.A.; Godinho, V.B.; Souza Riback, T.I.; Cruz, O.G.; Coelho, F.C. The expansion of Chikungunya in Brazil. *Lancet Reg. Health Am.* **2023**, *25*, 100571.

viruses

MDPI

Article

Comparison of Chikungunya Virus-Induced Disease Progression and Pathogenesis in Type-I Interferon Receptor-Deficient Mice (A129) and Two Wild-Type (129Sv/Ev and C57BL/6) Mouse Strains

Victoria A. Graham [1], Linda Easterbrook [1], Emma Rayner [1], Stephen Findlay-Wilson [1], Lucy Flett [1], Emma Kennedy [1], Susan Fotheringham [1], Sarah Kempster [2], Neil Almond [2] and Stuart Dowall [1,*]

[1] UK Health Security Agency (UKHSA), Porton Down, Salisbury SP4 0JG, Wiltshire, UK; victoria.graham@ukhsa.gov.uk (V.A.G.); linda.easterbrook@ukhsa.gov.uk (L.E.); emma.rayner@ukhsa.gov.uk (E.R.); stephen.findlay-wilson@ukhsa.gov.uk (S.F.-W.); lucy.flett@ukhsa.gov.uk (L.F.); emma.kennedy@ukhsa.gov.uk (E.K.); susan.fotheringham@ukhsa.gov.uk (S.F.)

[2] Medicines and Healthcare Products Regulatory Agency (MHRA), Blanche Ln, South Mimms, Potters Bar EN6 3QG, Hertfordshire, UK; sarah.kempster@mhra.gov.uk (S.K.); neil.almond@mhra.gov.uk (N.A.)

* Correspondence: stuart.dowall@ukhsa.gov.uk

Abstract: Chikungunya virus (CHIKV) is a mosquito-borne alphavirus causing a debilitating febrile illness with rheumatic disease symptoms of arthralgia and arthritis. Since its spread outside of Africa in 2005, it continues to cause outbreaks and disseminates into new territories. Intervention strategies are urgently required, including vaccination and antiviral approaches. To test efficacy, the use of small animal models is required. Two mouse strains, A129, with a deficiency in their type-I interferon (IFN) receptor, and C57BL/6 are widely used. A direct comparison of these strains alongside the wild-type parental strain of the A129 mice, 129Sv/Ev, was undertaken to assess clinical disease progression, viral loads in key tissues, histological changes and levels of sera biomarkers. Our results confirm the severe disease course in A129 mice which was not observed in the parental 129Sv/Ev strain. Of the two wild-type strains, viral loads were higher in 129Sv/Ev mice compared to C57BL/6 counterparts. Our results have established these models and parameters for the future testing of vaccines and antiviral approaches.

Keywords: Chikungunya; mouse; model; preclinical; pathogenesis

1. Introduction

Chikungunya virus (CHIKV) is an alphavirus within the *Togaviridae* family [1]. CHIKV was first isolated in 1952 from a human patient in Tanzania [2] with the virus being named after the Makonde word for "that which bends you up", in recognition of the severe joint pains associated with the disease [3].

For decades only local and occasional outbreaks were documented, until 2004 when CHIKV emerged at the coast of Kenya [4]. CHIKV is primarily transmitted by *Aedes aegypti* and *Aedes albopictus* mosquitoes, with a key mutation in the E1 gene (A226V) providing the virus with a gain-of-fitness adaptation enabling more efficient transmission by the latter species [5]. This likely contributed to the spread and magnitude of the outbreaks outside of mainland Africa in the Indian Ocean islands in 2005–2006, and subsequently across several countries in Southeast Asia [6,7]. In 2013, a CHIKV outbreak occurred on the Caribbean island of St. Martin [8] and thereafter, in 2014, spread to Brazil and across the American continent, resulting in more than 1.2 million cases in a single year [9]. *Aedes albopictus* are present in several areas of Europe [10], thus expanding the potential for the emergence of CHIKV into new geographical regions. This has been exemplified by the first autochthonous cases reported in Italy during 2007 [11] and France in 2010 [12].

CHIKV is listed as a priority pathogen by the UK Vaccine Network (UKVN) [13] and the Coalition for Epidemic Preparedness Innovations (CEPI) [14]; therefore, a healthy pipeline of vaccine candidates exist. One vaccine produced by Valneva has demonstrated seroprotective titres in a Phase 3 clinical trial [15], and has subsequently been approved by the US Food & Drug Administration (FDA) and European Medicines Agency (EMA). In addition, several antiviral compounds have been identified, but few have been assessed in animal models which is required to confirm their mode of action and effect [16]. The C57BL/6 mouse is wildly used in research due to substantial genetic homology with humans [17]. This strain has been used across multiple studies with CHIKV infection [18–25]. Similarly, A129 mice with a deficiency in their type-I interferon receptor (and similar knockout strains) are widely susceptible to viral disease, including CHIKV [20,22,23,26]. The wild-type parental strain, 129Sv/Ev, has been less used. This study therefore compared these three mouse strains after challenge with CHIKV, delivered subcutaneously near the foot to resemble natural infection via mosquito biting and feeding.

2. Materials and Methods

2.1. Virus

Chikungunya virus (strain LR 2006-OPY1) was kindly gifted from the Commissariat à l'Energie Atomique et aux Energies Alternatives (CEA) for use in this study. The virus originated from a French patient returning from La Reunion Island and passaged 3 times in Vero cells before a stock was produced on BHK-21 cells. Virus stock was titred on Vero cells to be 1.8×10^8 plaque-forming units/mL.

2.2. Animals

All experimental protocols with animals were undertaken according to the United Kingdom (UK) Animals (Scientific Procedures) Act 1986, with studies conducted under the authority of a UK Home Office approved project licence. The experimental protocols were approved by ethical review at UKHSA by the Animal Welfare and Ethical Review Body (AWERB; Approval Code: PPL P82D9CB4B). Female mice aged 5–8 weeks were obtained from UK Home Office approval suppliers: Marshall BioResources (strain A129 and 129Sv/Ev) and Envigo (C57BL/6). Mice were randomly allocated and housed in groups of 5–6 with food and water available *ad libitum* alongside regular environmental enrichment provided within cages. During and after challenge with CHIKV, all procedures, housing and husbandry took place inside a flexible film isolator housed within a Containment Level 3 facility. Prior to the start of the study, humane clinical endpoints were set which consisted of 20% weight loss compared to baseline; inactivity/immobility; neurological signs; or based on the advice of severe disease from the Named Animal Care and Welfare Officer (NACWO).

2.3. Study Design

To assess differences between the different mouse strains, according to the study groups, 5 mice were scheduled to be sampled at each timepoint based on power calculations (Wilcoxon and Mann–Whitney test, effect size 2.5, 95% significance and equal group allocation; G*Power 3.1.9.4). For survival analysis, 6 mice were assigned to each group based on a one-tailed Fisher's exact test (significance threshold of 0.05, 6 animals would give >80% power of demonstrating the difference to the control; G*Power 3.1.9.4). At day 3 and 7 post-challenge, groups were scheduled for cull to assess local responses between groups. The scheduled end of the study was day 7 for A129 mice and their 129Sv/Ev parental strain, and 14 days for the C57BL/6 strain. An overview of the study design is shown in Figure 1.

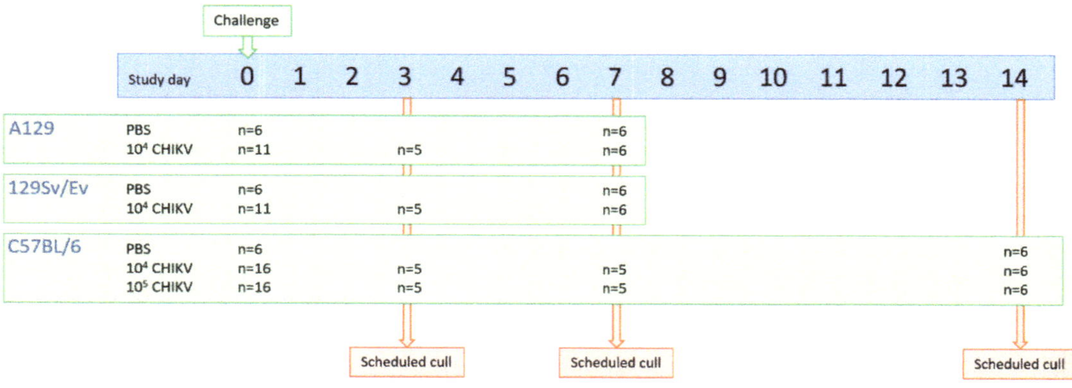

Figure 1. Schematic overview of study design assessing CHIKV infection in three mouse strains.

2.4. Challenge, Monitoring and Sampling

Mice were anaesthetised with isoflurane before being inoculated via the subcutaneous route with 40 µL of CHIKV or PBS control into the cranio-dorsal aspect of each distal hindlimb, just proximal to the tarsal joint. CHIKV was diluted with PBS to a ensure a concentration of either 10^4 or 10^5 pfu in the 80 µL inoculum distributed across the two limbs.

Body weight and temperature were monitored daily at the same time of day, the latter via an indwelling temperature chip (identiCHIP). Clinical and behavioural parameters were assessed and scored at least twice a day, with the frequency increasing to four times a day when observations considered to be moderate in severity were observed. Each observation was assigned a numerical value [1, eyes shut; 2, ruffled fur; 3, abnormal posture (hunched or arched), lethargy; 5, laboured breathing]; these were subsequently summed to derive a total cumulative score at each monitoring timepoint.

At necropsy, the right distal hindlimb below the stifle and samples of spleen and brain were placed into a PreCellys tube containing ceramic beads and stored at −80 °C for viral RNA assessment. The left distal hindlimb below the stifle and remainder of the spleen and brain were placed into 10% neutral-buffered formalin (NBF) for pathological examination. Blood was collected via cardiac puncture, with 100 µL added to animal RNAprotect blood tubes (Qiagen, Manchester, UK) and stored at −80 °C for viral RNA measurement. The remainder was placed into serum separation tubes (SST; Becton Dickinson, Wokingham, UK) with sera processed and stored at −80 °C for measurement of analytes by Luminex assay.

2.5. Viral RNA Measurement

Tissue samples for viral RNA analysis were weighed, resuspended in 1.5 mL PBS and homogenised using a PreCellys 24 homogeniser (Stretton Scientific, Alfreton, UK). Two hundred µL of tissue homogenate or blood was transferred to 600 µL RLT buffer (Qiagen, Manchester, UK) plus beta-mercaptoethanol and after at least 10 min, mixed with an equal volume of 70% ethanol. Tissues were further homogenised through a QIAshredder (Qiagen, Manchester, UK) at 16,000× g for 2 min and RNA extracted by KingFisher Flex automatic extraction using the BioSprint 96 one-for-all veterinary kit (Indical, Leipzig, Germany) as per manufacturer's instructions. RNA was eluted in 100 µL AVE buffer (Qiagen, Manchester, UK).

Samples were analysed by qRT-PCR using an assay that detects and quantifies CHIKV-specific amplicons via the one-step RT-PCR method [27]. The sensitivity of this method was determined using a series of 10-fold dilutions of in vitro transcribed RNA, generated from a cloned cDNA copy of the amplicon (pCH127), allowing readouts to be standardised to genome copies.

2.6. Histopathological Studies

The tissue samples fixed in 10% NBF were processed as follows: the hindlimb tarsal joint was transected sagittally and decalcified in 'Osteosoft' (Sigma Aldrich, Gillingham, UK) for 14 days. Together with sections of the spleen and brain, these were processed routinely into paraffin wax. Sections were cut to 4 μm and stained with haematoxylin and eosin (H&E). In addition, an in situ hybridisation technique ('RNAscope®') was used to detect the presence of CHIKV RNA. Briefly, tissues were pre-treated with hydrogen peroxide for 10 min (room temperature), followed by target retrieval for 15 min (98–101 °C) and protease plus for 30 min (40 °C) (Advanced Cell Diagnostics, Abingdon, UK). Tissues were then incubated with a CHIKV-specific probe (Catalogue no. 479508, Advanced Cell Diagnostics) for 2 h at 40 °C. Amplification of the signal was carried out following the RNAscope protocol using the RNAscope 2.5 HD Detection kit—Red (Advanced Cell Diagnostics).

Stained slides were scanned digitally using a 'Hamamatsu S360' digital slide scanner and examined using 'ndp.view2' software (v2.8.24) on a 4K digital monitor by a qualified pathologist who was blinded to the animal group details to minimise bias. Microscopic changes noted within the skeletal muscle, skin and subcutis were evaluated using a subjective scoring system that graded them as minimal, mild, moderate, or marked (for details of scoring methodology for skeletal muscle, skin and subcutis of the hindlimb samples, see Appendix A). Due to challenges encountered during microtomy, the joint tissues, including bone, were not present consistently in the sections and were therefore not scored. For the evaluation of the degree of staining for viral RNA, the following scoring system was used: 0 = no positive staining; 1 = minimal; 2 = mild; 3 = moderate; and 4 = abundant staining.

2.7. Luminex Analysis

A 32-plex mouse cytokine/chemokine panel was used consisting of granulocyte colony-stimulating factor (G-CSF), granulocyte-macrophage colony-stimulating factor (GM-CSF), eotaxin, interferon-gamma (IFNγ), interleukin(IL)-1a, IL-1β, IL-2, IL-3, IL-4, IL-5, IL-6, IL-7, IL-9, IL-10, IL-12(p40), IL-12(p70), IL-13, IL-15, IL-17, IFNγ-inducible protein 10 (IP-10), keratinocyte chemoattractant (KC), leukaemia inhibitory factor (LIF), lipopolysaccharide-induced CXC (LIX), macrophage colony-stimulating factor (M-CSF), macrophage inhibitory protein (MIP)-1α, MIP-1β, MIP-2, monocyte chemotactic protein 1 (MCP-1), monocyte induced by IFNγ (MIG), regulated upon activation, normal T cell expressed and presumably secreted (RANTES), tumor necrosis factor alpha (TNFα) and vascular endothelial growth factor (VEG-F) (Millipore, Watford, UK). The assay was performed according to the manufacturer's instructions. A standard preparation was diluted with assay buffer to produce a range covering concentrations of 10,000, 2000, 400, 80, 16 and 3.2 pg/mL. Within the Containment Level 3 (CL3) laboratory—required for handling samples from CHIKV-infected animals—25 μL of standard and quality control preparations or 25 μL sera were added to the relevant wells followed by 25 μL of serum matrix and 25 μL of premixed beads supplied with the kit. Plates were incubated at 4 °C overnight before being washed twice with 200 μL/well washing buffer using a handheld magnet. Following the wash steps, 25 μL of detector antibodies were added to all wells and the plate incubated for 1 h at room temperature. Next, 25 μL of streptavidin-phycoerythrin solution was added to each well and the plate was incubated for a further 30 min without any washing steps. After completion of staining, the microbeads were washed twice with wash buffer. To remove plates from the CL3 laboratory for analysis, beads were treated with formalin, previously shown to be amendable to this assay [28,29]. Beads were resuspended with 100 μL/well of 10% formalin solution made by dilution of 100% formalin (40% *w*/*v* formaldehyde solution) (Scientific Laboratory Supplies, Nottingham, UK) 1:9 *v*/*v* with phosphate buffered saline solution (Thermo Fisher, Loughborough, UK). Plates were fumigated with formaldehyde vapour overnight at room temperature for 16 h with the lids left ajar to allow vapour to reach all surfaces. Following fumigation, plates were placed into a sealed bag and removed from the CL3 laboratory.

Plates were washed twice with wash buffer, and once with sheath fluid in a Containment Level 2 laboratory to remove formalin solution before being resuspended in 150 µL of sheath fluid. Results were acquired on a Luminex MAGPIX instrument using Exponent software (v4.2.1324.1; Invitrogen, Paisley, UK). At least 50 events per region were collected and median fluorescence intensity (MFI) was measured. MFI values were converted to concentrations using results from the standard curve preparations.

2.8. Statistical Analysis

Statistical analyses were performed using MiniTab, v.16.2.2 (Minitab Inc, State College, PA, USA). A non-parametric Mann–Whitney statistical test was applied to ascertain significance between groups. A significance level below 0.05 was considered statistically significant.

3. Results

3.1. Time Course of Disease Progression

CHIKV uniformly caused severe disease in strain A129 mice lacking their type-I IFN receptor, with all challenged animals meeting humane endpoints by day 3, whereas wild-type mouse strains all survived until the scheduled end of the study (Figure 2A). In most groups, weights decreased at the start of the study irrespective of whether challenged with CHIKV or the PBS control, likely as a result of moving into different accommodation involving flexible film isolators required for biological containment. Weight soon after increased across groups, with the exception of a single A129 mouse which met humane clinical endpoint on day 3 and had lost 7.25% weight since the day of challenge (Figure 2B). Temperatures in A129 mice decreased post-challenge compared to the PBS group, a trend not observed with the 129Sv/Ev or C57BL/6 mouse strains (Figure 2C). Clinical signs of disease were observed only in the CHIKV-challenged A129 mice, consisting of ruffled fur, arched back, lethargy, eyes closed and laboured breathing (Figure 2D).

3.2. Viral Loads Measured in the Blood, Spleen, Limb and Brain at Day 3, 7 and 14 Post-Challenge

To ascertain the tropism and dissemination of CHIKV during infection, animals from each challenge group were scheduled for cull on days 3 and 7 post-challenge to assess tissue responses. Due to all A129 mice reaching humane clinical endpoints by day 3 post-challenge, samples were not available at later timepoints. Also, due to reaching humane endpoints and the sudden deterioration in health, only blood was collected from these animals for viral load analysis. All A129 mice had high levels of CHIKV RNA in their circulation, unlike in the wild-type mice with the exception of a single C57BL/6 animal at the day 14 timepoint (Figure 3A). The main site of viral RNA was in the lower limb, close to where the viral inoculum was administered (Figure 3B). The levels of viral RNA in the lower limb were significantly higher in 129Sv/Ev mice compared to C57BL/6 mice given the same challenge dose at day 3 post-challenge. Similarly, viral RNA was also observed in the spleen in the majority of 129Sv/Ev mice, whereas only sporadic and lower levels were detected in the C57BL/6 strain (Figure 3C) although these differences did not reach the significance threshold ($p = 0.0758$ and $p = 0.0601$ for days 3 and 7 post-challenge, respectively). For all samples tested, there were no statistically significant differences between the 10^4 and 10^5 challenge doses in the C57BL/6 mouse strain. No PCR responses were observed in the brain.

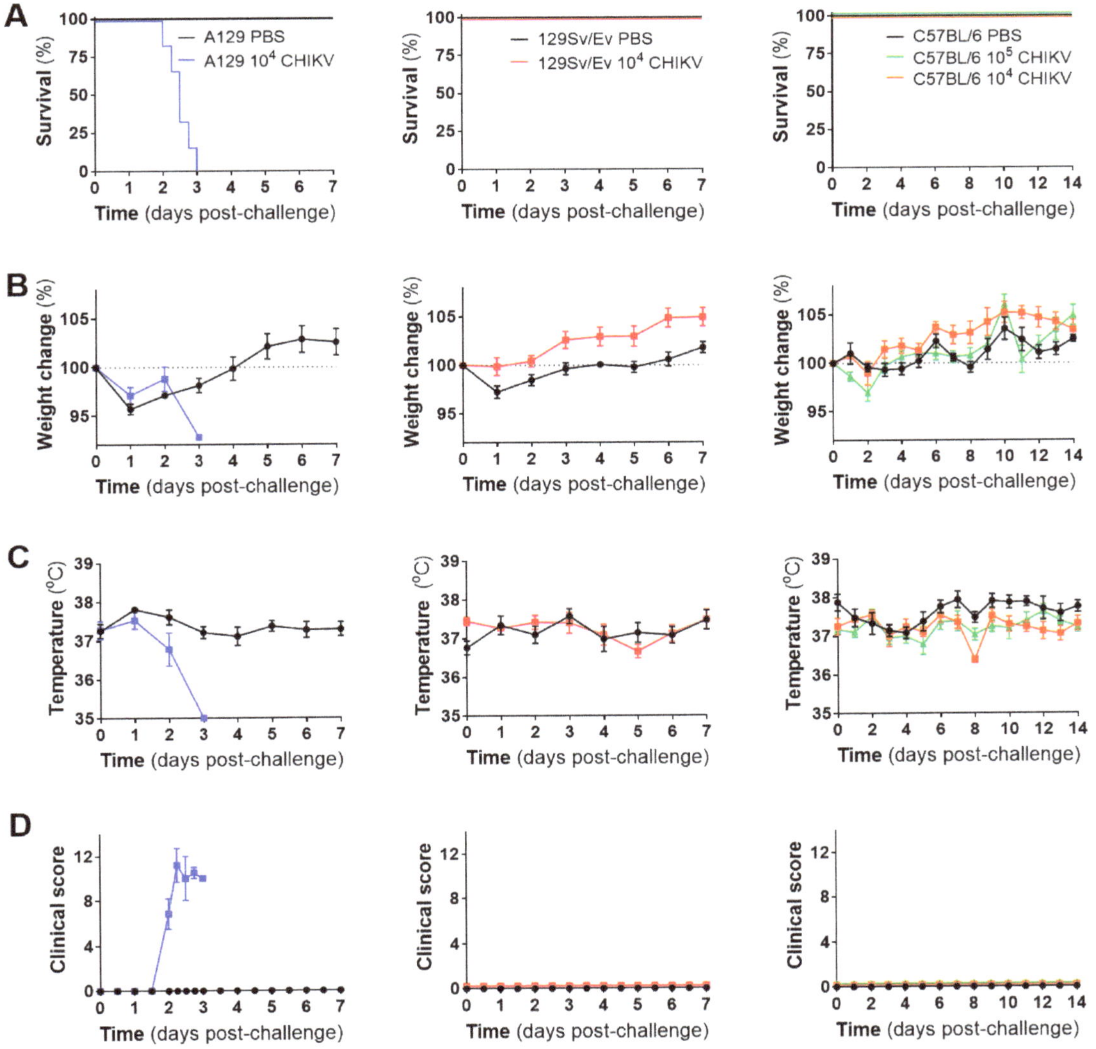

Figure 2. Clinical outcomes of mice challenged with CHIKV. (**A**) Kaplan–Meier survival plots, (**B**) weight changes compared to the day of challenge, (**C**) body temperature and (**D**) clinical scores. Lines show mean values with error bars denoting standard error. Results are shown from n = 6 animals per group pre-allocated for monitoring of clinical course for up to 7 days (A129 and 129Sv/Ev) or 14 days (C57BL/6) post-challenge.

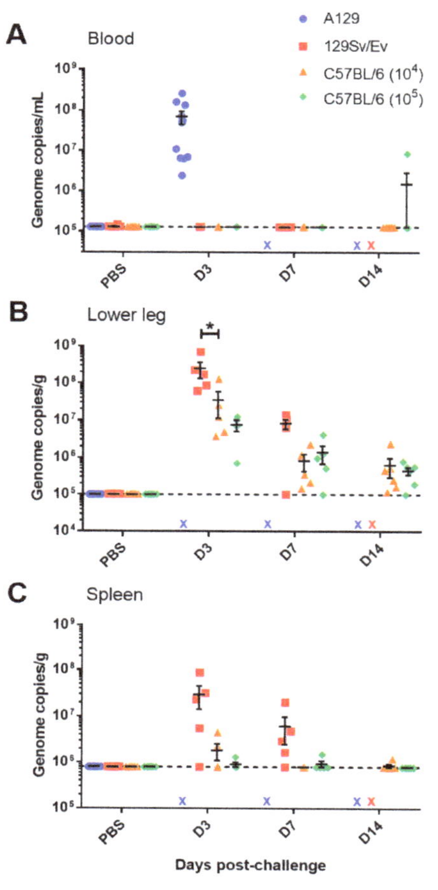

Figure 3. Viral RNA levels in blood and tissues from CHIKV-challenged mice measured by RT-PCR. Levels in the (**A**) blood expressed as genome copies per mL, and levels in the (**B**) lower hindlimb and (**C**) spleen expressed as genome copies per g. x, indicates samples not available. *, $p < 0.05$. Symbols show results from individual animals with line and whisker plots denoting mean and standard error.

3.3. Pathological Findings

Microscopic changes associated with infection with the CHIK virus was observed in a proportion of animals from each of the challenged groups at the various timepoints (Figures 4 and 5). In the skeletal muscle, individual or groups of myocytes were observed to be degenerate or necrotic, with irregular-shaped cellular outlines and absent cytoplasmic cross striations, floccular, hyper-eosinophilic cytoplasm and nuclear karyorrhexis (coagulation necrosis). Variable numbers of mononuclear inflammatory cells, primarily macrophages, infiltrated the cells or were within the epimysium. In other areas, inflammatory cells infiltrated the endo- and peri- and epimysium of intact myocytes. Myocyte regeneration was also observed variably, comprising reduced diameter fibres with multiple, centralised nuclei and basophilic cytoplasm. Inflammatory cells extended into surrounding fibro-vascular and adipose connective tissue; this was variably associated with pale, eosinophilic material expanding collagenous tissue (oedema), vascular congestion and variable erythrocyte extravasation (haemorrhage).

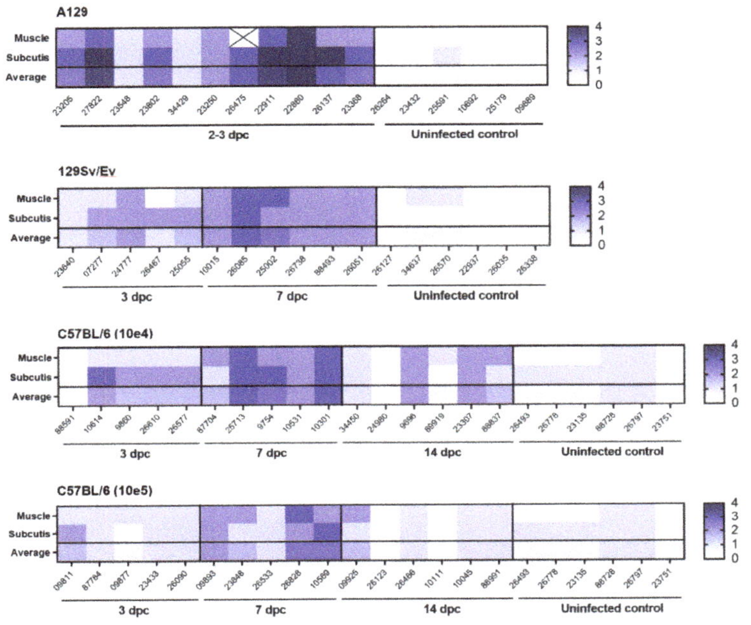

Figure 4. Heatmap illustrating the severity of microscopic changes in muscle and subcutaneous tissue (subcutis) of the hindlimb in individual animals from mouse strains A129, 129SvEv and C57BL/6. Values indicate individual and average severity scores (muscle and subcutis). Cross symbol indicates sample not available.

These changes were noted with the greatest severity in soft tissues associated with the tarsal joint of all A129 mice at days 2–3 post-challenge. In the other groups at the day 3 timepoint, changes were present in five/five animals in the 129Sv/Ev group, four/five animals in the C57BL/6 group challenged with 10^4 pfu CHIKV, and five/five animals of the same strain challenged with 10^5 pfu CHIKV. Appreciable differences in the severity of these changes between these three groups were not prominent.

By contrast, by day 7 post-challenge, there was a trend towards increased severity of changes compared to the previous timepoint, the extent of which appeared comparable between the three groups. Only the C57BL/6 strain animals were examined at 14 days post-challenge; overall, the severity of disease had reduced and were similar between the two dosing regimens. Minimal changes were noted infrequently in the control groups that received a PBS inoculation.

Microscopic changes consistent with infection with CHIKV were not observed in either the spleen or brain of any animal in any group.

Staining for CHIK viral RNA was present variably in a proportion of animals where microscopic changes had been identified (Figure 6). As expected, the degree of staining was greatest in tissues of animals in the A129 group at 2–3 days post-challenge, with severity scores ranging from 3 to 4 in all animals. At this timepoint, staining was also noted in four/five animals in the 129Sv/Ev group (severity scores 1–2); in three/five C57BL/6 animals that received the 10^4 dose (severity scores 1–2); and in two/five animals that received the 10^5 dose (severity scores 1–2). By day 7 post-challenge, positive staining was present in four/six 129Sv/Ev strain animals (severity scores 1–2), two/five C57BL/6 animals that received the 10^4 dose (severity score 1); and in one/five animals that received the 10^5 dose (severity score 1). In the C57BL/6 strain groups at 14 days post-challenge, viral staining was absent in all animals at both dose groups. Staining for viral RNA was absent in all control groups.

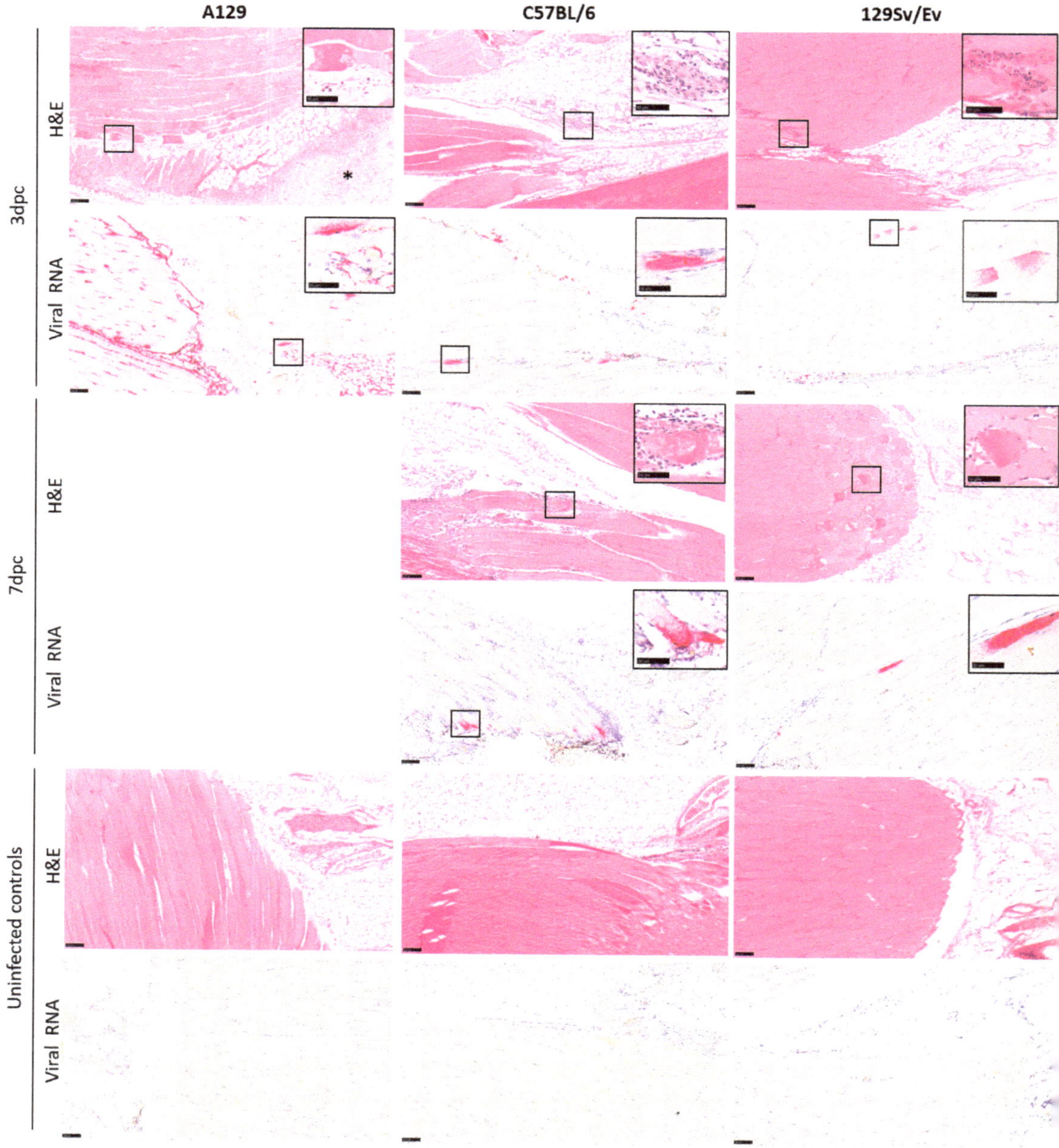

Figure 5. Representative images illustrating the type and severity of microscopic changes, and the presence of viral RNA staining, in the skeletal muscle and subcutis of the hindlimb from A129, C57BL/6 and 129Sv/Ev mouse strains at 3 and 7 days after challenge with 10^4 pfu CHIKV, alongside uninfected controls. The changes comprise skeletal myocyte degeneration and loss, variably associated with a mainly neutrophilic cell infiltration and inflammation of the subcutis with oedema and haemorrhage and concomitant infiltrating inflammatory cells (indicated by an asterisk). These changes are more

apparent in the A129 mice at 3 days post-challenge (left column, row 1) compared to C57BL/6 and 129Sv/Ev mouse strains at the same timepoint (middle and right columns, row 1). Viral staining is noted in all three strains (left, middle and right columns, row 2), and most prominent in A129 mice (left column, row 2). At 7 days post-challenge, there is a comparable increase in severity of changes in the C57BL/6 and 129Sv/Ev mouse strains (middle and right columns, rows 3 and 4) and low-level viral staining (middle and right columns, rows 4 and 5). Microscopic and changes and viral staining are absent in the uninfected control animals (left, middle and right columns, rows 5 and 6). Inset, higher power images of changes are highlighted in square boxes. Scale bars represent 100 µm in the main images and 50 µm in the insets. H&E, ISH.

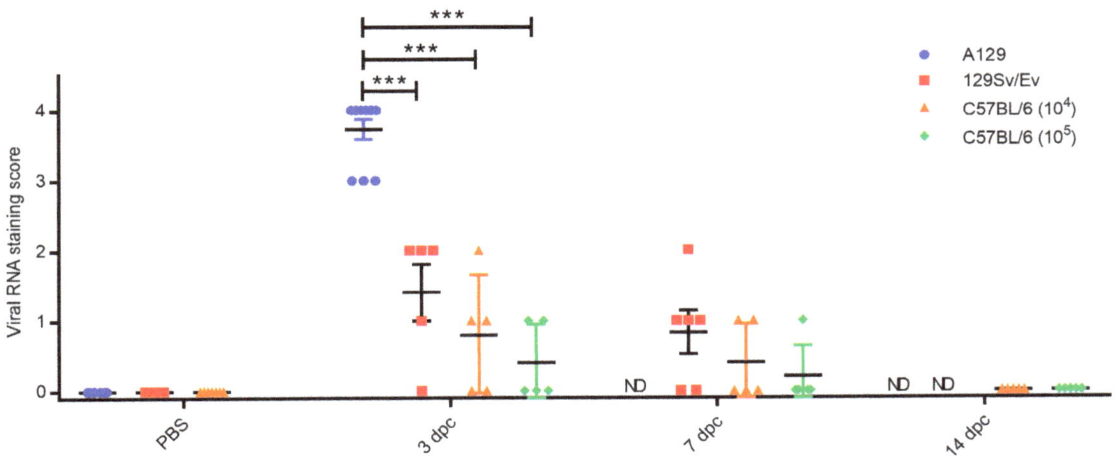

Figure 6. Levels of RNAscope staining in the muscle and subcutis of the hindlimb. PBS group are unchallenged animals whereas other groups were challenged with CHIKV and culled at 3, 7 and 14 days post-challenge (dpc). ND, not done due to animals meeting humane endpoints beforehand (A129) or not planned as part of the study schedule. Symbols show results from individual animals with line and whisker plots denoting mean and standard error. ***, $p < 0.001$.

3.4. Cytokine, Chemokine and Growth Factor Levels Associated with CHIKV Infection

Concentrations of 32 analytes were measured in the sera of animals challenged with CHIKV and compared with levels of those of the PBS control group. Statistical analysis showed that the majority of levels were significantly decreased during the acute phases of infection, with only levels of IFN-γ, IP-10 and MIG showing increased levels (Figure 7). The peak changes for the 129Sv/Ev mice were observed at 3 days post-challenge, whereas for the C57BL/6 strain it was at 7 days, with the latter showing a greater number of analytes affected at this timepoint with the higher challenge group.

One cytokine (IFN-γ) and three chemokines (IP-10, KC and MIG) showing significant differences were further analysed. Levels of IFN-γ increased in 129Sv/Ev mice, reaching significance at the day 7 timepoint, in contrast to stable levels observed in C57BL/6 mice (Figure 8A). IP-10 levels were again clearly increasing in 129Sv/Ev mice, but with the C57BL/6 mice, although statistically significant differences were detected, the increases in levels were small (Figure 8B). The baseline levels of KC were higher in C57BL/6 mice compared with the 129Sv/Ev strain, with significant decreases observed in the former (Figure 8C). By day 14, KC levels had returned to normal in those challenged with 10^5, but still remained subdued in those challenged with 10^4. For MIG, the levels in all 129Sv/Ev-challenged mice at day 7 were higher than in the PBS group, with levels remaining consistent amongst other groups (Figure 8D).

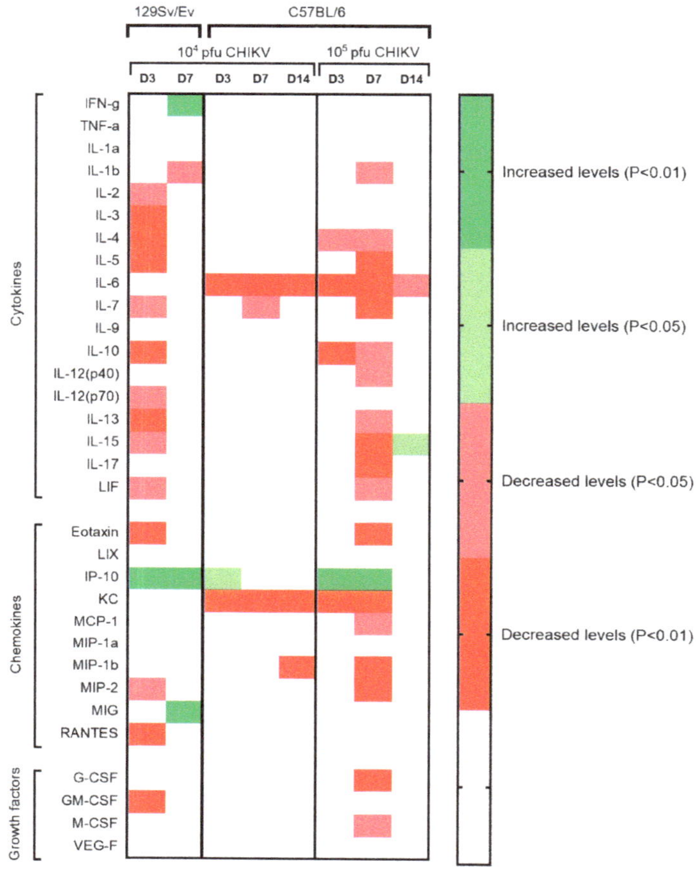

Figure 7. Heatmap showing statistically significant changes in cytokine, chemokine and growth factor serum concentrations in CHIKV-challenged animals compared to PBS mock-challenged strain-matched control animals (n = 6). Uncoloured areas represent no significance observed ($p > 0.05$).

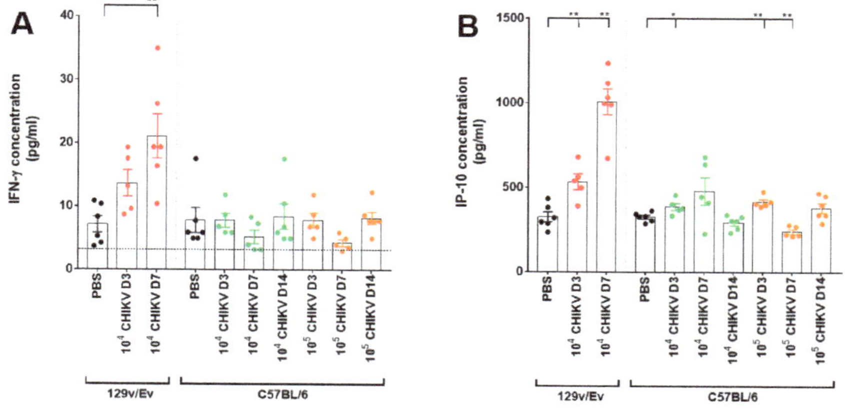

Figure 8. *Cont.*

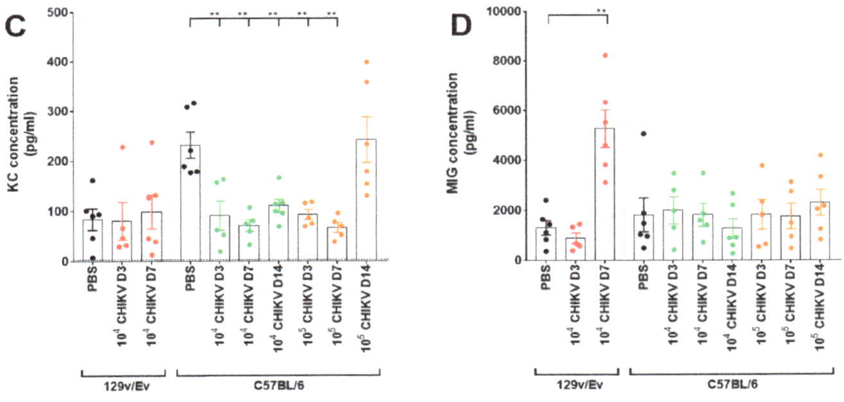

Figure 8. Concentrations of a subset of analytes demonstrating significant differences after challenge with CHIKV. Results show levels of (**A**) IFN-γ, (**B**) IP-10, (**C**) KC and (**D**) MIG. Bars represent the mean values with error bars denoting standard error and results from individual animals identified as coloured dots. *, $p < 0.05$; **, $p < 0.01$.

4. Discussion

Our results provide a direct comparison of CHIKV pathogenesis by applying identical study protocols in three mouse strains that are widely used for modelling arbovirus infection. The CHIKV strain used, LR 2006-OPY1, was isolated from a traveller returning from La Réunion in 2006 [30,31], where during this outbreak, over 260,000 cases (approximately one-third of the population) were reported [32]. This strain has also been widely used in challenge studies of non-human primates [33–37], thus our studies provide a direct comparison between these different preclinical model systems.

Direct comparisons were made between the two wild-type mouse strains, C57BL/6 and 129Sv/Ev. The results presented corroborate findings that C57BL/6 mice are a robust model for studying the pathogenesis of CHIKV-induced musculoskeletal disease [19]. Interestingly, in 129Sv/Ev mice, the levels of viral RNA in the lower leg and spleen were higher compared with the C57BL/6 strain, yet there was no evidence of viremia. The finding of viral RNA in the spleen indicates dissemination, and evidence of viral material 7 days post-challenge contrasts with reports that virus in this strain of mouse is cleared within 5 days [38]. However, our assay relied on the RT-PCR detection of viral RNA which may be more sensitive than live virus assays but has the limitation of not being able to distinguish between a viable virus and non-viable genetic nucleic acid. Our results provide evidence that 129Sv/Ev mice are a useful alternative to C57BL/6 mice, with increased and more prolonged virus in local tissues. In another comparison with CHIKV infection in wild-type mouse strains, CD-1 mice have also been shown to be comparable with the C57BL/6 strain [39].

Alongside the wild-type mouse strains, type-I IFN receptor-deficient mice were also assessed. All A129 mice met humane endpoints by day 3, slightly earlier than another study where survival dropped from day 4 [38]. This may be due to interpretation of endpoints, where in our study, humane clinical endpoints were set before initiation of the study whereas in the latter, the study was reported as a morbidity/mortality design. The advantage of A129 mice is the ability to measure survival as a clear outcome post-challenge, as has been demonstrated with CHIKV vaccines based on chimpanzee adenovirus vectored constructs [40,41]. The A129 mice showed consistent viremia compared with the wild-type mouse strains, as seen in humans and at similar levels to the 7 $\log_{10}$ magnitude reported [42]. Due to a rapid deterioration in health and multiple animals reaching humane endpoint simultaneously, tissues were not able to be sampled at necropsy. The finding of viremia demonstrates further utility of this strain to be used for vector competence studies, as reported for the similar AG129 mice with an additional knockout for their IFN-γ receptor [43].

Across the different mouse strains, a consistent challenge dose of 10^4 pfu was delivered to enable comparability. This dose was used in an assessment of different CHIKV strains, including LR 2006 OPY1 used in our study, to assess lineage-specific differences in virulence using the A129 mouse model after footpad inoculation with all strains resulting in animals meeting humane endpoints [44]. In other reports using the same CHIKV-challenge strain as used herein, a vaccine study using Balb/C mice similarly challenged with 10^4 pfu by the subcutaneous (s.c.) route [45]. In C57BL/6 mice, 10^4 pfu was also used, but with the s.c site being at the base of the tail [46]. Higher doses of 10^6 $TCID_{50}$ in C57BL/6 mice have been reported, resulting in a transient and recoverable disease progression [47]. Whilst $TCID_{50}$ titres are not directly comparable to pfu, an efficacy study in C57BL/6 mice used a challenge dose of 10^5 pfu via the s.c. footpad route [48]. Infectious clones of LR 2006 OPY1 were used to develop a challenge model in C57BL/6 mice, with doses of 10^4, 10^5 and 10^6 pfu all demonstrating similar levels of viraemia with no discernible differences observed [49]. Given these higher doses of CHIKV strain LR 2006 OPY1 in C57BL/6 mice, this provided the reasoning for inclusion of an additional group of 10^5 pfu in this mouse strain.

Foot swelling has been observed with CHIKV infection in mice but was not observed in the studies described herein. Some studies have been conducted in younger C57BL/6 mice, such as those 14 days old [19], whereas others were in mice more aligned with being over 6 weeks old [18,50]. Our findings of failing to detect swelling aligns with previous reports conducted with wild-type 129Sv/Ev mice where no swelling of the contralateral hind footpad was detectable [38]. In C57BL/6J mice, there was no change in limb thickness in animals challenged with 10^2 and 10^4 pfu, but it was increased with a higher dose of 10^6 pfu using an Indian CHIKV isolate [51]. After inoculation at the base of the tail, foot swelling was also absent despite the presence of viraemia, alongside the finding that non-injected feet do not demonstrate clinical effects [18]. Therefore, the lack of swelling may be due to minor differences in the injection sites, as the foot/footpad is used in many reports [18,19], but due to welfare considerations and discomfort of the animals post-inoculation, a site slightly above the foot was used in our studies. This argument is reinforced by others, who found that delivery of CHIKV via the flank of C57BL/6 mice did not result in detectable foot swelling, but when delivered to the feet, swelling was observed [49]. Mosquito-mediated human infection results in swelling across multiple joints [52], and not just those directly where the biting occurred, thus presenting a limitation of the C57BL/6 mouse model as these effects are not recapitulated. Interestingly, when this was tested in BALB/c mice, after inoculation on the back, swelling and inflammation of the legs was observed [53], indicating differences between mouse strains but this was more likely due to the young age (2–3 days) of the mice used which are more susceptible to alphaviral disease [54].

The type of histopathological lesions observed in the skeletal muscle and subcutis associated with the tarsal joint of the C57BL/6 strain of mice are similar with published data [18,19,50,55]; unfortunately, due to limitations in the processing of tissues, the joints and bone were not included in the evaluation. Similar types of pathological changes were noted in the other two mouse strains with varying severity, with the most severe pathology, as predicted, noted in the A129, type-I IFN receptor-deficient mice. The severity of changes observed in the three mice strains at the varying timepoints were consistent with the changes seen in the viral and clinical data.

At set timepoints post-challenge, planned culls were scheduled to ascertain local responses at days 3 and 7 post-challenge based on evidence from published studies. Day 3 was chosen due to viremia at days 2 and 3 being higher than day 1 in A129 mice [44,45]. In addition, days 3 and 7 also correlated with the biphasic disease progression in the wild-type C57BL/6 mice, where peak disease/morbidity scores were observed on day 3 in two independent studies, and then another at day 7 or soon after [47,50]. With the aim of our work being to compare the different mouse strains (A129, 129Sv/Ev and C57BL/6), we chose representative timepoints relevant to these. For further interrogation of the A129 mouse strain, due to the rapid progression of disease, earlier timepoints specific to this strain are warranted and for the wild-type strains, and additional later timepoints should be considered.

During early infection events, levels of the chemokine IP-10 were consistently increased in CHIKV-challenged 129Sv/Ev and C57BL/6 mice. MIG and IFN-γ were also elevated, but only in 129Sv/Ev mice. These align with studies in acute CHIKV patients. In a study of 196 acute patients across tertiary care hospitals across India, levels of IP-10, IFN-γ and MIG sera levels correlated with disease severity [56], although a fourth marker which was also found to be significant, MCP-1, did not show correlation in our challenged mice. In 88 CHIKV patients in Brazil, levels of circulating IP-10 and IFN-γ were increased during the acute phase [57], further aligning with our findings in CHIKV-challenged mice; although other markers were also increased in the patients such as TNF-α, IL-7 and MCP-1 which were not significant in our mouse models. Another study in Brazil comparing 29 acute CHIKV patients with 21 healthy controls also found significantly increased levels of IP-10, MIG and IFN-γ in the patient sera, alongside IFN-α, IL-6, IL-8, IL-10 and MCP-1 [58]. Similarly, in acutely infected CHIKV patients, levels of IP-10 were increased after studying levels from 69 patient from the Gabonese outbreak of 2007, but again, alongside several other analytes [59]. IFN-γ has been shown to be elevated in similar studies in C57BL/6 mice, but alongside other biomarkers including TNF-α, MCP-1, IL-6 and IFN-α/β [18].

The majority of other cytokines were either substantially reduced or unchanged compared with control groups. A meta-analysis of biomarkers in CHIKV infection found that decreased levels of RANTES and IL-8 had an association with disease severity; however, the rise of a significant number of analytes (including those included in the mouse panel: IL-6, TNF-α, MCP-1, IL-12, IL-13, GM-CSF, IL-1 α, IL-15, IL-10, IL-4, MIP-1α, and MIP-1 β) were associated with chronification [60]. The levels in mice being decreased for many of these analytes suggests that the infection in these mice is resolving instead of progressing to a chronic state. The finding of fewer analytes being raised in our studies may arise from differences in levels of sensitivities between different assays used or degradation of cytokines, due to their intrinsically short half-life in blood [61]. Due to decontamination and logistical processes from collection at necropsy to processing the sera in an in vitro laboratory and frozen storage, delays may have resulted in the deterioration of the quality of the sample. However, in previous work on Zika virus, a similar approach was undertaken with a wider range of biomarkers showing elevated levels [62], thus this is unlikely an effect due to the condition of the sera being tested.

In summary, our results point towards two separate CHIKV-induced disease sequelae in mice which should be measured in studies on the impact of vaccines and therapeutics. The A129 mouse model, for which CHIKV produces a uniformly lethal disease, allows survival readouts as a relevant endpoint for success criteria of intervention testing. Of the wild-type mice strains tested, in the absence of clinical disease, viral loads and histological changes would be appropriate readouts. In that regard, viral loads were higher in 129Sv/Ev compared with C57BL/6 in key sites; thus, the former strain may prove more sensitive when incomplete protection is conferred by the intervention and would be preferable for deciphering partial control of virus replication and other protective effects.

Given the lack of feasibility to conduct efficacy studies due to the unpredictable nature of CHIKV outbreaks and incomplete epidemiology, in part due to inadequate surveillance [63], animal models form a key component for vaccine licensure. The use of immunogenicity endpoints as a surrogate of protection, based on the passive transfer of human post-vaccination sera to a susceptible animal model (e.g., NHP) prior to wild-type challenge to ascertain protective effects [64] in the licensure of the first human CHIKV vaccine, demonstrates the value of in vivo models which resemble the spectrum of human disease manifestations.

Author Contributions: Conceptualization, V.A.G., S.K., N.A. and S.D.; methodology, V.A.G., L.E., E.R., S.F.-W., L.F., E.K. and S.F.; formal analysis, E.R.; investigation, V.A.G., S.F.-W., S.F. and S.D.; writing—original draft preparation, S.D.; writing—review and editing, V.A.G., E.R. and S.K.; funding acquisition, S.K. and N.A. All authors have read and agreed to the published version of the manuscript.

Funding: This research was funded by Innovate UK and the Department of Health and Social Care (Project title "Serological Vaccine Standards for Emerging Diseases"; File Ref. 971613, awarded to MHRA) and is funded through Official Development Assistance (ODA).

Institutional Review Board Statement: The animal study protocol was approved through ethical review by the Animal Welfare and Ethical Review Body (AWERB) at the UK Health Security Agency (UKHSA) and its predecessor organisations.

Informed Consent Statement: Not applicable.

Data Availability Statement: The data presented in this study are available upon request from the corresponding author.

Acknowledgments: The authors gratefully acknowledge the support from the Biological Investigations Group, Virology & Pathogenesis Group and Pathology Group at UKHSA, Porton. The views expressed in this article are those of the authors and not necessarily those of the employing institutes or the funding bodies.

Conflicts of Interest: The authors declare no conflicts of interest. The funders had no role in the design of the study; in the collection, analyses, or interpretation of data; in the writing of the manuscript; or in the decision to publish the results.

Appendix A. Scoring Criteria for the Subjective Assessment of Microscopic Changes in Hindlimb Soft Tissues Resulting from Infection with Chikungunya Virus in Mice

Lesion	Score				
	0 (Normal)	1 (Minimal)	2 (Mild)	3 (Moderate)	4 (Marked)
Skeletal muscle					
Skeletal myocyte degeneration/necrosis ± inflammatory cell infiltration, mainly neutrophils.	None	Occasional, scattered single or small groups of degenerating myocytes—up to 5% of skeletal muscle affected.	Increased numbers of degenerating myocytes—6–25% of skeletal muscle affected.	Frequent numbers of degenerating myocytes—26–50% of skeletal muscle affected.	Numerous degenerating myocytes—>50% of skeletal muscle affected.
Skin and subcutis					
Vascular congestion, oedema and haemorrhage with a mixed inflammatory cell infiltrate, often predominantly neutrophils, with fewer macrophages and lymphocytes.	None	Up to 5% of skin and soft connective tissues affected.	6–25% of skin and connective tissues affected.	26–50% of skin and connective tissues affected.	>50% of skin and connective tissues affected.

References

1. Strauss, J.H.; Strauss, E.G. The alphaviruses: Gene expression, replication, and evolution. *Microbiol. Rev.* **1994**, *58*, 491–562. [CrossRef]
2. Lumsden, W.H. An epidemic of virus disease in Southern Province, Tanganyika Territory, in 1952-53. II. General description and epidemiology. *Trans. R. Soc. Trop. Med. Hyg.* **1955**, *49*, 33–57. [CrossRef] [PubMed]
3. Robinson, M.C. An epidemic of virus disease in Southern Province, Tanganyika Territory, in 1952-53. I. Clinical features. *Trans. R. Soc. Trop. Med. Hyg.* **1955**, *49*, 28–32. [CrossRef]

4. Chretien, J.P.; Anyamba, A.; Bedno, S.A.; Breiman, R.F.; Sang, R.; Sergon, K.; Powers, A.M.; Onyango, C.O.; Small, J.; Tucker, C.J.; et al. Drought-associated chikungunya emergence along coastal East Africa. *Am. J. Trop. Med. Hyg.* **2007**, *76*, 405–407. [CrossRef] [PubMed]
5. Tsetsarkin, K.A.; Vanlandingham, D.L.; McGee, C.E.; Higgs, S. A single mutation in chikungunya virus affects vector specificity and epidemic potential. *PLoS Pathog.* **2007**, *3*, e201. [CrossRef] [PubMed]
6. Kumarasamy, V.; Prathapa, S.; Zuridah, H.; Chem, Y.K.; Norizah, I.; Chua, K.B. Re-emergence of Chikungunya virus in Malaysia. *Med. J. Malays.* **2006**, *61*, 221–225.
7. Laras, K.; Sukri, N.C.; Larasati, R.P.; Bangs, M.J.; Kosim, R.; Djauzi; Wandra, T.; Master, J.; Kosasih, H.; Hartati, S.; et al. Tracking the re-emergence of epidemic chikungunya virus in Indonesia. *Trans. R. Soc. Trop. Med. Hyg.* **2005**, *99*, 128–141. [CrossRef]
8. Cassadou, S.; Boucau, S.; Petit-Sinturel, M.; Huc, P.; Leparc-Goffart, I.; Ledrans, M. Emergence of chikungunya fever on the French side of Saint Martin island, October to December 2013. *Euro Surveill.* **2014**, *19*, 20752. [CrossRef]
9. Zeller, H.; Van Bortel, W.; Sudre, B. Chikungunya: Its History in Africa and Asia and Its Spread to New Regions in 2013-2014. *J. Infect. Dis.* **2016**, *214*, S436–S440. [CrossRef]
10. Medlock, J.M.; Hansford, K.M.; Schaffner, F.; Versteirt, V.; Hendrickx, G.; Zeller, H.; Van Bortel, W. A review of the invasive mosquitoes in Europe: Ecology, public health risks, and control options. *Vector Borne Zoonotic Dis.* **2012**, *12*, 435–447. [CrossRef]
11. Angelini, R.; Finarelli, A.C.; Angelini, P.; Po, C.; Petropulacos, K.; Macini, P.; Fiorentini, C.; Fortuna, C.; Venturi, G.; Romi, R.; et al. An outbreak of chikungunya fever in the province of Ravenna, Italy. *Euro Surveill.* **2007**, *12*, 3260. [CrossRef] [PubMed]
12. Grandadam, M.; Caro, V.; Plumet, S.; Thiberge, J.M.; Souares, Y.; Failloux, A.B.; Tolou, H.J.; Budelot, M.; Cosserat, D.; Leparc-Goffart, I.; et al. Chikungunya virus, southeastern France. *Emerg. Infect. Dis.* **2011**, *17*, 910–913. [CrossRef] [PubMed]
13. Noad, R.J.; Simpson, K.; Fooks, A.R.; Hewson, R.; Gilbert, S.C.; Stevens, M.P.; Hosie, M.J.; Prior, J.; Kinsey, A.M.; Entrican, G.; et al. UK vaccines network: Mapping priority pathogens of epidemic potential and vaccine pipeline developments. *Vaccine* **2019**, *37*, 6241–6247. [CrossRef] [PubMed]
14. Gouglas, D.; Christodoulou, M.; Plotkin, S.A.; Hatchett, R. CEPI: Driving Progress toward Epidemic Preparedness and Response. *Epidemiol. Rev.* **2019**, *41*, 28–33. [CrossRef]
15. Schneider, M.; Narciso-Abraham, M.; Hadl, S.; McMahon, R.; Toepfer, S.; Fuchs, U.; Hochreiter, R.; Bitzer, A.; Kosulin, K.; Larcher-Senn, J.; et al. Safety and immunogenicity of a single-shot live-attenuated chikungunya vaccine: A double-blind, multicentre, randomised, placebo-controlled, phase 3 trial. *Lancet* **2023**, *401*, 2138–2147. [CrossRef]
16. Battisti, V.; Urban, E.; Langer, T. Antivirals against the Chikungunya Virus. *Viruses* **2021**, *13*, 1307. [CrossRef]
17. Nguyen, D.; Xu, T. The expanding role of mouse genetics for understanding human biology and disease. *Dis. Model. Mech.* **2008**, *1*, 56–66. [CrossRef]
18. Gardner, J.; Anraku, I.; Le, T.T.; Larcher, T.; Major, L.; Roques, P.; Schroder, W.A.; Higgs, S.; Suhrbier, A. Chikungunya virus arthritis in adult wild-type mice. *J. Virol.* **2010**, *84*, 8021–8032. [CrossRef]
19. Morrison, T.E.; Oko, L.; Montgomery, S.A.; Whitmore, A.C.; Lotstein, A.R.; Gunn, B.M.; Elmore, S.A.; Heise, M.T. A mouse model of chikungunya virus-induced musculoskeletal inflammatory disease: Evidence of arthritis, tenosynovitis, myositis, and persistence. *Am. J. Pathol.* **2011**, *178*, 32–40. [CrossRef]
20. Schilte, C.; Couderc, T.; Chretien, F.; Sourisseau, M.; Gangneux, N.; Guivel-Benhassine, F.; Kraxner, A.; Tschopp, J.; Higgs, S.; Michault, A.; et al. Type I IFN controls chikungunya virus via its action on nonhematopoietic cells. *J. Exp. Med.* **2010**, *207*, 429–442. [CrossRef]
21. Wang, E.; Volkova, E.; Adams, A.P.; Forrester, N.; Xiao, S.Y.; Frolov, I.; Weaver, S.C. Chimeric alphavirus vaccine candidates for chikungunya. *Vaccine* **2008**, *26*, 5030–5039. [CrossRef] [PubMed]
22. Werneke, S.W.; Schilte, C.; Rohatgi, A.; Monte, K.J.; Michault, A.; Arenzana-Seisdedos, F.; Vanlandingham, D.L.; Higgs, S.; Fontanet, A.; Albert, M.L.; et al. ISG15 is critical in the control of Chikungunya virus infection independent of UbE1L mediated conjugation. *PLoS Pathog.* **2011**, *7*, e1002322. [CrossRef]
23. Couderc, T.; Chretien, F.; Schilte, C.; Disson, O.; Brigitte, M.; Guivel-Benhassine, F.; Touret, Y.; Barau, G.; Cayet, N.; Schuffenecker, I.; et al. A mouse model for Chikungunya: Young age and inefficient type-I interferon signaling are risk factors for severe disease. *PLoS Pathog.* **2008**, *4*, e29. [CrossRef]
24. Teo, T.H.; Lum, F.M.; Claser, C.; Lulla, V.; Lulla, A.; Merits, A.; Renia, L.; Ng, L.F. A pathogenic role for CD4+ T cells during Chikungunya virus infection in mice. *J. Immunol.* **2013**, *190*, 259–269. [CrossRef]
25. Schilte, C.; Buckwalter, M.R.; Laird, M.E.; Diamond, M.S.; Schwartz, O.; Albert, M.L. Cutting edge: Independent roles for IRF-3 and IRF-7 in hematopoietic and nonhematopoietic cells during host response to Chikungunya infection. *J. Immunol.* **2012**, *188*, 2967–2971. [CrossRef] [PubMed]
26. Partidos, C.D.; Weger, J.; Brewoo, J.; Seymour, R.; Borland, E.M.; Ledermann, J.P.; Powers, A.M.; Weaver, S.C.; Stinchcomb, D.T.; Osorio, J.E. Probing the attenuation and protective efficacy of a candidate chikungunya virus vaccine in mice with compromised interferon (IFN) signaling. *Vaccine* **2011**, *29*, 3067–3073. [CrossRef]
27. Edwards, C.J.; Welch, S.R.; Chamberlain, J.; Hewson, R.; Tolley, H.; Cane, P.A.; Lloyd, G. Molecular diagnosis and analysis of Chikungunya virus. *J. Clin. Virol.* **2007**, *39*, 271–275. [CrossRef] [PubMed]
28. Dowall, S.D.; Graham, V.A.; Fletcher, T.; Hewson, R. Use and reliability of multiplex bead-based assays (Luminex) at Containment Level 4. *Methods* **2019**, *158*, 17–21. [CrossRef]

29. Dowall, S.D.; Graham, V.A.; Tipton, T.R.; Hewson, R. Multiplex cytokine profiling with highly pathogenic material: Use of formalin solution in luminex analysis. *J. Immunol. Methods* **2009**, *348*, 30–35. [CrossRef]
30. Tsetsarkin, K.; Higgs, S.; McGee, C.E.; De Lamballerie, X.; Charrel, R.N.; Vanlandingham, D.L. Infectious clones of Chikungunya virus (La Reunion isolate) for vector competence studies. *Vector Borne Zoonotic Dis.* **2006**, *6*, 325–337. [CrossRef]
31. Parola, P.; de Lamballerie, X.; Jourdan, J.; Rovery, C.; Vaillant, V.; Minodier, P.; Brouqui, P.; Flahault, A.; Raoult, D.; Charrel, R.N. Novel chikungunya virus variant in travelers returning from Indian Ocean islands. *Emerg. Infect. Dis.* **2006**, *12*, 1493–1499. [CrossRef] [PubMed]
32. Pialoux, G.; Gauzere, B.A.; Jaureguiberry, S.; Strobel, M. Chikungunya, an epidemic arbovirosis. *Lancet Infect. Dis.* **2007**, *7*, 319–327. [CrossRef] [PubMed]
33. Akahata, W.; Yang, Z.Y.; Andersen, H.; Sun, S.; Holdaway, H.A.; Kong, W.P.; Lewis, M.G.; Higgs, S.; Rossmann, M.G.; Rao, S.; et al. A virus-like particle vaccine for epidemic Chikungunya virus protects nonhuman primates against infection. *Nat. Med.* **2010**, *16*, 334–338. [CrossRef] [PubMed]
34. Erasmus, J.H.; Auguste, A.J.; Kaelber, J.T.; Luo, H.; Rossi, S.L.; Fenton, K.; Leal, G.; Kim, D.Y.; Chiu, W.; Wang, T.; et al. A chikungunya fever vaccine utilizing an insect-specific virus platform. *Nat. Med.* **2017**, *23*, 192–199. [CrossRef]
35. Roques, P.; Ljungberg, K.; Kummerer, B.M.; Gosse, L.; Dereuddre-Bosquet, N.; Tchitchek, N.; Hallengard, D.; Garcia-Arriaza, J.; Meinke, A.; Esteban, M.; et al. Attenuated and vectored vaccines protect nonhuman primates against Chikungunya virus. *JCI Insight* **2017**, *2*, e83527. [CrossRef]
36. Roy, C.J.; Adams, A.P.; Wang, E.; Plante, K.; Gorchakov, R.; Seymour, R.L.; Vinet-Oliphant, H.; Weaver, S.C. Chikungunya vaccine candidate is highly attenuated and protects nonhuman primates against telemetrically monitored disease following a single dose. *J. Infect. Dis.* **2014**, *209*, 1891–1899. [CrossRef]
37. Labadie, K.; Larcher, T.; Joubert, C.; Mannioui, A.; Delache, B.; Brochard, P.; Guigand, L.; Dubreil, L.; Lebon, P.; Verrier, B.; et al. Chikungunya disease in nonhuman primates involves long-term viral persistence in macrophages. *J. Clin. Investig.* **2010**, *120*, 894–906. [CrossRef]
38. Gardner, C.L.; Burke, C.W.; Higgs, S.T.; Klimstra, W.B.; Ryman, K.D. Interferon-alpha/beta deficiency greatly exacerbates arthritogenic disease in mice infected with wild-type chikungunya virus but not with the cell culture-adapted live-attenuated 181/25 vaccine candidate. *Virology* **2012**, *425*, 103–112. [CrossRef]
39. Anderson, E.J.; Knight, A.C.; Heise, M.T.; Baxter, V.K. Effect of Viral Strain and Host Age on Clinical Disease and Viral Replication in Immunocompetent Mouse Models of Chikungunya Encephalomyelitis. *Viruses* **2023**, *15*, 1057. [CrossRef]
40. Campos, R.K.; Preciado-Llanes, L.; Azar, S.R.; Lopez-Camacho, C.; Reyes-Sandoval, A.; Rossi, S.L. A Single and Un-Adjuvanted Dose of a Chimpanzee Adenovirus-Vectored Vaccine against Chikungunya Virus Fully Protects Mice from Lethal Disease. *Pathogens* **2019**, *8*, 231. [CrossRef]
41. Campos, R.K.; Preciado-Llanes, L.; Azar, S.R.; Kim, Y.C.; Brandon, O.; Lopez-Camacho, C.; Reyes-Sandoval, A.; Rossi, S.L. Adenoviral-Vectored Mayaro and Chikungunya Virus Vaccine Candidates Afford Partial Cross-Protection from Lethal Challenge in A129 Mouse Model. *Front. Immunol.* **2020**, *11*, 591885. [CrossRef] [PubMed]
42. Weaver, S.C.; Osorio, J.E.; Livengood, J.A.; Chen, R.; Stinchcomb, D.T. Chikungunya virus and prospects for a vaccine. *Expert Rev. Vaccines* **2012**, *11*, 1087–1101. [CrossRef] [PubMed]
43. Baldon, L.V.R.; de Mendonca, S.F.; Ferreira, F.V.; Rezende, F.O.; Amadou, S.C.G.; Leite, T.; Rocha, M.N.; Marques, J.T.; Moreira, L.A.; Ferreira, A.G.A. AG129 Mice as a Comprehensive Model for the Experimental Assessment of Mosquito Vector Competence for Arboviruses. *Pathogens* **2022**, *11*, 879. [CrossRef]
44. Langsjoen, R.M.; Haller, S.L.; Roy, C.J.; Vinet-Oliphant, H.; Bergren, N.A.; Erasmus, J.H.; Livengood, J.A.; Powell, T.D.; Weaver, S.C.; Rossi, S.L. Chikungunya Virus Strains Show Lineage-Specific Variations in Virulence and Cross-Protective Ability in Murine and Nonhuman Primate Models. *mBio* **2018**, *9*, e02449-17. [CrossRef]
45. Chattopadhyay, A.; Aguilar, P.V.; Bopp, N.E.; Yarovinsky, T.O.; Weaver, S.C.; Rose, J.K. A recombinant virus vaccine that protects against both Chikungunya and Zika virus infections. *Vaccine* **2018**, *36*, 3894–3900. [CrossRef]
46. Rao, S.; Abeyratne, E.; Freitas, J.R.; Yang, C.; Tharmarajah, K.; Mostafavi, H.; Liu, X.; Zaman, M.; Mahalingam, S.; Zaid, A.; et al. A booster regime of liposome-delivered live-attenuated CHIKV vaccine RNA genome protects against chikungunya virus disease in mice. *Vaccine* **2023**, *41*, 3976–3988. [CrossRef]
47. Teo, T.H.; Her, Z.; Tan, J.J.; Lum, F.M.; Lee, W.W.; Chan, Y.H.; Ong, R.Y.; Kam, Y.W.; Leparc-Goffart, I.; Gallian, P.; et al. Caribbean and La Reunion Chikungunya Virus Isolates Differ in Their Capacity to Induce Proinflammatory Th1 and NK Cell Responses and Acute Joint Pathology. *J. Virol.* **2015**, *89*, 7955–7969. [CrossRef] [PubMed]
48. Voigt, E.A.; Fuerte-Stone, J.; Granger, B.; Archer, J.; Van Hoeven, N. Live-attenuated RNA hybrid vaccine technology provides single-dose protection against Chikungunya virus. *Mol. Ther.* **2021**, *29*, 2782–2793. [CrossRef]
49. Hallengard, D.; Kakoulidou, M.; Lulla, A.; Kummerer, B.M.; Johansson, D.X.; Mutso, M.; Lulla, V.; Fazakerley, J.K.; Roques, P.; Le Grand, R.; et al. Novel attenuated Chikungunya vaccine candidates elicit protective immunity in C57BL/6 mice. *J. Virol.* **2014**, *88*, 2858–2866. [CrossRef]
50. Goupil, B.A.; McNulty, M.A.; Martin, M.J.; McCracken, M.K.; Christofferson, R.C.; Mores, C.N. Novel Lesions of Bones and Joints Associated with Chikungunya Virus Infection in Two Mouse Models of Disease: New Insights into Disease Pathogenesis. *PLoS ONE* **2016**, *11*, e0155243. [CrossRef]

51. Jain, J.; Narayanan, V.; Kumar, A.; Shrinet, J.; Srivastava, P.; Chaturvedi, S.; Sunil, S. Establishment and Comparison of Pathogenicity and Related Neurotropism in Two Age Groups of Immune Competent Mice, C57BL/6J Using an Indian Isolate of Chikungunya Virus (CHIKV). *Viruses* **2019**, *11*, 578. [CrossRef] [PubMed]
52. Ng, K.W.; Chow, A.; Win, M.K.; Dimatatac, F.; Neo, H.Y.; Lye, D.C.; Leo, Y.S. Clinical features and epidemiology of chikungunya infection in Singapore. *Singap. Med. J.* **2009**, *50*, 785–790.
53. Dhanwani, R.; Khan, M.; Lomash, V.; Rao, P.V.; Ly, H.; Parida, M. Characterization of chikungunya virus induced host response in a mouse model of viral myositis. *PLoS ONE* **2014**, *9*, e92813. [CrossRef]
54. Seymour, R.L.; Adams, A.P.; Leal, G.; Alcorn, M.D.; Weaver, S.C. A Rodent Model of Chikungunya Virus Infection in RAG1 -/- Mice, with Features of Persistence, for Vaccine Safety Evaluation. *PLoS Negl. Trop. Dis.* **2015**, *9*, e0003800. [CrossRef]
55. Chang, A.Y.; Tritsch, S.R.; Porzucek, A.J.; Schwartz, A.M.; Seyler-Schmidt, M.; Glass, A.; Latham, P.S.; Reid, S.P.; Simon, G.L.; Mores, C.N. A Mouse Model for Studying Post-Acute Arthritis of Chikungunya. *Microorganisms* **2021**, *9*, 1998. [CrossRef]
56. Babu, N.; Mahilkar, S.; Jayaram, A.; Ibemgbo, S.A.; Mathur, G.; Shetty, U.; Sudandiradas, R.; Kumar, P.S.; Singh, S.; Pani, S.S.; et al. Cytokine profile, neutralisation potential and viral replication dynamics in sera of chikungunya patients in India: A cross-sectional study. *Lancet Reg. Health Southeast Asia* **2023**, *19*, 100269. [CrossRef] [PubMed]
57. Alves de Souza, T.M.; Fernandes-Santos, C.; Araujo da Paixao de Oliveira, J.; Tome, L.C.T.; Fiestas-Solorzano, V.E.; Nunes, P.C.G.; Guimaraes, G.M.C.; Sanchez-Arcila, J.C.; Paiva, I.A.; de Souza, L.J.; et al. Increased Indoleamine 2,3-Dioxygenase 1 (IDO-1) Activity and Inflammatory Responses during Chikungunya Virus Infection. *Pathogens* **2022**, *11*, 444. [CrossRef] [PubMed]
58. Tanabe, I.S.B.; Santos, E.C.; Tanabe, E.L.L.; Souza, S.J.M.; Santos, F.E.F.; Taniele-Silva, J.; Ferro, J.F.G.; Lima, M.C.; Moura, A.A.; Anderson, L.; et al. Cytokines and chemokines triggered by Chikungunya virus infection in human patients during the very early acute phase. *Trans. R. Soc. Trop. Med. Hyg.* **2019**, *113*, 730–733. [CrossRef]
59. Wauquier, N.; Becquart, P.; Nkoghe, D.; Padilla, C.; Ndjoyi-Mbiguino, A.; Leroy, E.M. The acute phase of Chikungunya virus infection in humans is associated with strong innate immunity and T CD8 cell activation. *J. Infect. Dis.* **2011**, *204*, 115–123. [CrossRef]
60. Ferreira, A.S.; Baldoni, N.R.; Cardoso, C.S.; Oliveira, C.D.L. Biomarkers of severity and chronification in chikungunya fever: A systematic review and meta-analysis. *Rev. Inst. Med. Trop. Sao Paulo* **2021**, *63*, e16. [CrossRef]
61. Deckers, J.; Anbergen, T.; Hokke, A.M.; de Dreu, A.; Schrijver, D.P.; de Bruin, K.; Toner, Y.C.; Beldman, T.J.; Spangler, J.B.; de Greef, T.F.A.; et al. Engineering cytokine therapeutics. *Nat. Rev. Bioeng.* **2023**, *1*, 286–303. [CrossRef] [PubMed]
62. Dowall, S.D.; Graham, V.A.; Hewson, R. Lineage-dependent differences of Zika virus infection in a susceptible mouse model are associated with different profiles of cytokines, chemokines, growth factors and acute phase proteins. *Cytokine* **2020**, *125*, 154864. [CrossRef] [PubMed]
63. Bettis, A.A.; L'Azou Jackson, M.; Yoon, I.K.; Breugelmans, J.G.; Goios, A.; Gubler, D.J.; Powers, A.M. The global epidemiology of chikungunya from 1999 to 2020: A systematic literature review to inform the development and introduction of vaccines. *PLoS Negl. Trop. Dis.* **2022**, *16*, e0010069. [CrossRef] [PubMed]
64. Roques, P.; Fritzer, A.; Dereuddre-Bosquet, N.; Wressnigg, N.; Hochreiter, R.; Bossevot, L.; Pascal, Q.; Guehenneux, F.; Bitzer, A.; Corbic Ramljak, I.; et al. Effectiveness of CHIKV vaccine VLA1553 demonstrated by passive transfer of human sera. *JCI Insight* **2022**, *7*, e160173. [CrossRef] [PubMed]

viruses

MDPI

Review

The Evolving Role of Zika Virus Envelope Protein in Viral Entry and Pathogenesis

Ashkan Roozitalab [1], Jiantao Zhang [1], Chenyu Zhang [1], Qiyi Tang [2] and Richard Y. Zhao [1,3,4,5,6,*]

1 Department of Pathology, University of Maryland School of Medicine, Baltimore, MD 21201, USA; aroozitalab@som.umaryland.edu (A.R.); chenyu.zhang@som.umaryland.edu (C.Z.)
2 Department of Microbiology, Howard University College of Medicine, Washington, DC 20059, USA; qiyi.tang@Howard.edu
3 Department of Microbiology and Immunology, University of Maryland School of Medicine, Baltimore, MD 21201, USA
4 Institute of Global Health, University of Maryland School of Medicine, Baltimore, MD 21201, USA
5 Institute of Human Virology, University of Maryland School of Medicine, Baltimore, MD 21201, USA
6 Research and Development Service, VA Maryland Heath Care System, Baltimore, MD 21201, USA
* Correspondence: rzhao@som.umaryland.edu; Tel.: +1-410-706-6301

Abstract: Zika virus (ZIKV) was first discovered in Uganda's Zika Forest in 1947. The early African viruses posed little or no health risk to humans. Since then, ZIKV has undergone extensive genetic evolution and adapted to humans, and it now causes a range of human diseases, including neurologically related diseases in adults and congenital malformations such as microcephaly in newborns. This raises a critical question as to why ZIKV has become pathogenic to humans, and what virological changes have taken place and enabled it to cause these diseases? This review aims to address these questions. Specifically, we focus on the ZIKV envelope (E) protein, which is essential for initiating infection and plays a crucial role in viral entry. We compare various virologic attributes of E protein between the ancestral African strains, which presumably did not cause human diseases, with epidemic strains responsible for current human pathogenesis. First, we review the role of the ZIKV E protein in viral entry and endocytosis during the viral life cycle. We will then examine how the E protein interacts with host immune responses and evades host antiviral responses. Additionally, we will analyze key differences in the sequence, structure, and post-translational modifications between African and Asian lineages, and discuss their potential impacts on viral infection and pathogenesis. Finally, we will evaluate neutralizing antibodies, small molecule inhibitors, and natural compounds that target the E protein. This will provide insights into the development of potential vaccines and antiviral therapies to prevent or treat ZIKV infections and associated diseases.

Keywords: Zika virus; envelope protein; viral entry; microcephaly; inhibitors; neutralizing antibodies; antibody-dependent enhancement; ancestral and epidemic viral strains; African and Asian lineages

1. Introduction

1.1. Zika Virus (ZIKV): A Brief History

The Zika virus (ZIKV) is an enveloped, positive-sense, single-stranded RNA (+ssRNA) virus belonging to the *orthoflavivirus* genus within the family of Flaviviridae, which includes other medically significant arboviruses such as West Nile virus (WNV), Dengue virus (DENV), Japanese Encephalitis virus (JEV), and Chikungunya virus (CHIKV) [1]. These viruses are transmitted to humans and other animals through the bite of infected arthropods, such as mosquitoes *Aedes aegyptis*; thus, they are referred to as "arboviruses", short for "arthropod-borne viruses". Arboviruses cause a range of illnesses, from mild to severe, with serious health consequences. For example, WNV is prevalent worldwide and is the most common arbovirus in the U.S., potentially leading to severe neurological diseases such as encephalitis and meningitis, although many infections remain asymptomatic.

ZIKV was first isolated from a rhesus monkey in Uganda's Zika Forest in 1947 [2]. Early studies showed that infected monkeys displayed mild or no symptoms, but mice younger than two weeks were highly susceptible to intraperitoneal (i.p.) inoculation; whereas older mice rarely became infected, likely due to their fully developed blood-brain barriers (BBB), which prevented ZIKV from accessing the brain. Subsequent research confirmed ZIKV's ability to infect the central nervous system (CNS) in mice [3].

Although ZIKV is primarily transmitted by *Aedes* mosquitoes to both humans and animals, it initially appeared less harmful to humans [4]. The first human ZIKV infection was recorded in 1952 [5], and since then, ZIKV antibodies have been detected in human sera from various countries [6], although no severe diseases were initially linked to these infections. ZIKV can also infect humans via non-vector-borne routes, including sexual transmission, maternal-fetal transmission, and blood transfusions [7–9]. Most individuals infected with ZIKV are asymptomatic or experience mild symptoms such as rash, fever, joint pain, and conjunctivitis.

The first laboratory strain of ZIKV, MR766 ($ZIKV_{MR766}$), was named after the febrile rhesus monkey number 766 from which it was first isolated in 1947 in the Zika Forest of Uganda, during studies of yellow fever [5,10]. This strain, now referred to as the ancestral or historical ZIKV strain, has been widely used for research into the virus's biology, pathogenesis, and vaccine development. $ZIKV_{MR766}$ demonstrated high neurotropism in mice, with the virus being recoverable only from infected mouse brains [4].

Since its initial discovery, ZIKV has spread globally beyond Africa, with cases reported in Asia and the Americas [1,6]. Over the past seventy plus years, ZIKV has migrated from Africa to Asia, across the Pacific, and eventually to the Americas. Early ZIKV infections, primarily in Africa and Asia, were not linked to severe disease, and its spread eastward from Africa included countries such as Pakistan, Malaysia, and Indonesia [11–13]. Outbreaks occurred in these regions during 1977–1978, but they were relatively small. However, as the virus crossed the Pacific Ocean, significant outbreaks occurred on Pacific islands, including Yap Island in Micronesia in 2007 and in French Polynesia in 2013–2014. These outbreaks were noteworthy due to the increased incidence of Guillain–Barré syndrome-like symptoms and, in some cases, birth defects [9,14]. The virus reached the Americas in 2015, with a major outbreak starting in Brazil. From there, ZIKV spread rapidly across Central and South America, and the Caribbean. In 2016, the World Health Organization (WHO) declared ZIKV a global public health emergency [15]. During this outbreak, ZIKV infections were reported in eighty-five countries, territories, or subnational areas, affecting an estimated 1.5 million individuals globally, according to the WHO. Brazil had the hardest hit, with estimates ranging from 440,000 to 1.3 million cases by December 2015 [16]. Although the spread of ZIKV has declined since 2017, the virus remains a global threat, with the potential for significant outbreaks again. A recent example is the 2021 outbreak in Kerala, India, during the second wave of the COVID-19 pandemic [17]. As of May 2024, ZIKV transmission had been reported in ninety-two countries and territories, according to the WHO. The history of ZIKV epidemiology and evolution clearly demonstrates that the virus has gradually adapted to humans as a host. Over nearly eighty years, changes within the viral genome have driven the emergence of more pathogenic ZIKV strains, contributing to its increasing threat to human health [1,18].

1.2. The Structure and Genomic Organization of ZIKV

The structure and morphology of ZIKV closely resemble those of other orthoflaviviruses, such as DENV [19]. Specifically, the viral surface proteins are arranged in an icosahedral-like symmetry. ZIKV particles measure approximately 40–65 nm in diameter, with surface projections ranging from 5–10 nm [19], resulting in an average overall size of 45–75 nm. As an enveloped virus, ZIKV is surrounded by a lipid envelope derived from the host cell membrane. Inside this envelope, the viral RNA genome is encased within a protein capsid, approximately 25–30 nm in diameter. This capsid is surrounded by a lipid bilayer primarily composed of three structural proteins: the envelope (E) protein, the membrane (M) protein, and the capsid (C) protein. The E protein covers most of the virion surface, organized into "rafts" of three E protein dimers lying parallel to each other. Beneath the E protein lies the M protein, which plays a crucial role in viral maturation. The C protein forms the shell that encloses the viral RNA genome (Figure 1A).

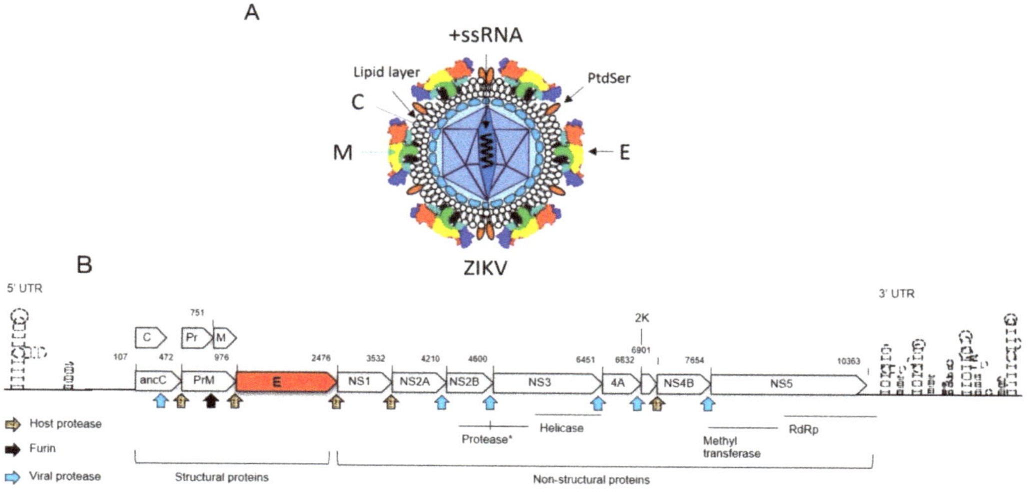

Figure 1. Schematic representation of the ZIKV viral particle and genome. (**A**) The Zika viral particle is surrounded by a lipid envelope. The E protein, organized into "rafts" of three E protein dimers lying parallel to each other, covers most of the virion surface. Beneath the E protein lies the M protein. The C protein forms the shell that encloses the viral +ssRNA genome. Phosphatidylserine (PtdSer), a phospholipid, is interlocked with E proteins to form "raft-like" structures on the viral surface. (**B**) This diagram illustrates the genomic organization of ZIKV, with each viral protein depicted according to its relative position within the RNA genome. Protease cleavage sites are indicated by arrows, representing cleavage by the viral protease, host protease, and furin protease. The numbers above each protein product denote their start and end positions within the genome. The abbreviations used are as follows: anaC (anchored capsid protein C), C (capsid protein C), prM (precursor membrane protein), M (membrane protein), Pr (protein pr), E (envelope protein), NS (nonstructural protein), 2K (signal peptide 2K), and UTR (untranslated region). The viral protease consists of the N-terminal domain of NS3 and the C-terminal domain of NS2B, as described in the text (*). The C-terminal region of NS3 encodes a helicase, while NS5 contains a methyltransferase (MTase) at its N-terminal end and an RNA-dependent RNA polymerase (RdRp) at its C-terminal end. This diagram is adapted from [1].

ZIKV's genome is +ssRNA of approximately 10.79 kb in size (Figure 1B) [1,19]. It contains a single open reading frame (ORF) flanked by two non-coding regions (NCRs): the 5′ NCR (107 nucleotides) and the 3′ NCR (428 nucleotides) [1,20]. The ZIKV genome encodes a polyprotein that undergoes post-translational processing to produce three structural proteins (C, prM, and E) and seven non-structural proteins (NS1, NS2A, NS2B, NS3, NS4A, NS4B, and NS5), which are necessary for viral replication and assembly [19,21,22]. The non-structural proteins are processed to yield four viral enzymes: proteases (PR), helicase, methyltransferase (MTase), and RNA-dependent RNA polymerase (RdRP). These enzymes play essential roles in viral genome replication, assembly, and evasion of host immune defenses [19]. The structural proteins (C, prM, and E) are essential for the formation of the viral capsid and envelope. The ZIKV C protein plays a crucial role in packaging the viral RNA genome and assembling new virions [23]. In addition, The ZIKV C protein also functions as a viral suppressor of RNA interference by directly interacting with and inhibiting the endoribonuclease activity of host Dicer enzymes in human neural stem cells (NSCs) [24]. The mature C protein is generated when the viral PR cleaves the anchor-C (anaC) protein. This cleavage also triggers the splitting of the precursor membrane (prM) protein, resulting in the mature M protein, and a Pr protein [25–27]. Furin cleavage of prM into M in the trans-Golgi network (TGN) before release of mature virions leads to the formation of mature, infectious viral particles [28,29]. The M protein stabilizes the E protein by locking it into a fixed mature conformation, thereby preventing premature membrane fusion during transit through the secretory pathway [30]. Research findings demonstrated that a single amino acid (a.a.) change (S139N) in the ZIKV prM protein, found in most epidemic strains, significantly increases neurovirulence in neonatal mice and induces greater cell death in human neural progenitor cells (hNPCs) compared to the N139S mutant or pre-epidemic strains [23,24]. The E protein plays a key role in binding to host cell receptors and mediating fusion with the host cell membrane during viral entry and endocytosis [31]. ZIKV non-structural proteins carry out essential functions in viral replication, immune evasion, and pathogenesis. NS1 is crucial for viral replication and helps the virus evade the host immune system [32]. The NS2B-NS3 protease is essential for ZIKV replication and maturation, as NS2B acts as a cofactor enabling NS3 to cleave the viral polyprotein into functional non-structural proteins [33]. Functional studies have demonstrated that NS4A and NS4B modulate host immune responses and are associated with neurodevelopmental damage [34]. NS5, the largest ZIKV protein, contains both RdRp and MTase domains, essential for viral RNA synthesis and capping [32].

1.3. The Viral Infection Cycle

To initiate infection, ZIKV first binds to host cell surface receptors, such as AXL, TYRO3, and MER, which are members of the TAM (TYRO3, AXL, and MER) receptor tyrosine kinase family, a process facilitated by the E protein (Figure 2). The virus then enters host cells via clathrin-mediated endocytosis, where it fuses with the endosomal membrane in a pH-dependent manner. This viral–host membrane fusion releases the viral genomic RNA into the cytoplasm. The +ssRNA of ZIKV acts directly as messenger RNA (mRNA), allowing immediate translation upon entering the host cell. The translation of the viral genome produces a polyprotein, which undergoes enzymatic cleavage to generate both structural and non-structural proteins as described above. Viral replication occurs on the surface of the endoplasmic reticulum (ER), where new viral RNA strands are synthesized. The viral RdRP synthesizes double-stranded RNA (dsRNA) from the positive-sense RNA genome. This dsRNA is then transcribed and replicated to produce additional viral genomes. Electron microscopy (EM) studies have shown that ZIKV replication is confined to the ER compartment in astroglial cells and neurons, indicating membrane-associated viral replication within the CNS [35]. During pregnancy, ZIKV replicates in the placenta and fetal brain, particularly in the Hofbauer cells of the placenta [36], which can serve as a reservoir for viral replication and dissemination of the virus to the fetus, leading to potential complications like microcephaly. Immature virus particles are assembled in the ER and then pass through the TGN, where they mature fully. Mature virions bud from the host cell membrane, acquiring a new viral envelope before being released into the extracellular environment [1,37]. Once released, these new viral particles can infect neighboring cells, initiating another round of the infectious cycle [38].

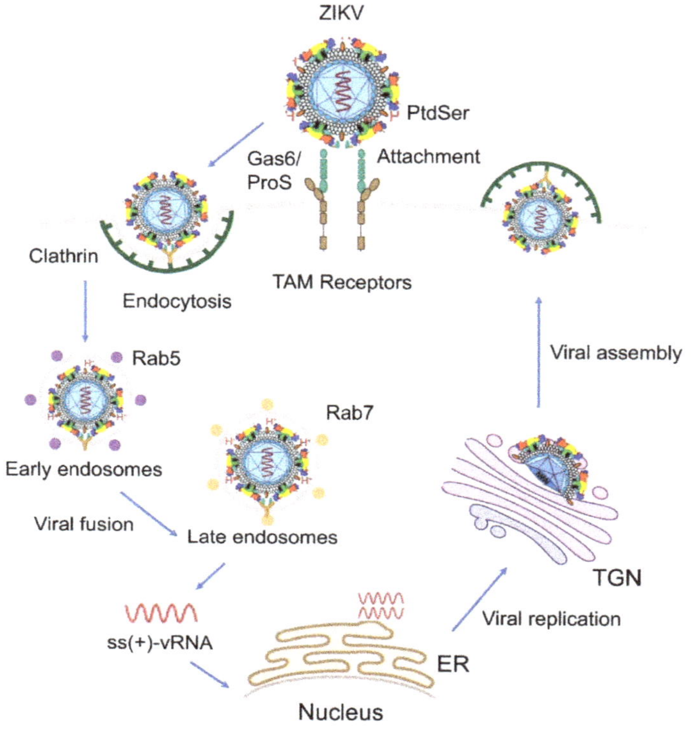

Figure 2. Schematic overview of the ZIKV life cycle. Upon contact with host cells, ZIKV initiates infection by binding to TAM family receptors on the host cell surface via the bridging molecules Growth arrest-specific 6 (Gas6) and Protein S (ProS), a process mediated by the viral E protein. Note that, besides TAM receptors, other receptors can also be used, which are described in the text. Following attachment, the virus enters host cells through clathrin-mediated endocytosis. Within endosomes, the acidic environment triggers fusion between the viral envelope and the endosomal membrane, leading to the release of the single-stranded, positive-sense viral RNA [ss (+) vRNA] into the cytoplasm. The viral RNA is immediately translated to produce viral proteins. Replication of the viral genome occurs on the ER surface, where double-stranded RNA (dsRNA) intermediates are synthesized and subsequently used to generate additional viral genomes. Immature virions are assembled in the ER, transported through the TGN for maturation, and finally released from the host cell as fully infectious particles. For indications of different color structures on ZIKV, please see Figure 1A.

2. ZIKV Envelope Protein

ZIKV E protein is the primary structural component present on the surface of the viral particle, forming a layer of rafts composed of three parallel E protein dimers. This protein plays a pivotal role in the viral life cycle, facilitating viral binding to host cell receptors, mediating membrane fusion during viral entry, and interacting with the host's innate and adaptive immune responses. The E protein is crucial for the virus's ability to infect cells and trigger pathogenic effects [1,20,39–41].

2.1. The Structure of ZIKV E Protein

The E protein exists as a dimer on the surface of the mature virion ZIKV. The monomer of E protein is composed of 504 amino acids (a.a.) and has a molecular weight of approximately 54 kDa. Structurally, it is divided into four key domains: three N-terminal ectodomains (DI, DII, and DIII; 1–406 a.a.) and a C-terminal stem-transmembrane (TM) domain (407–504 a.a.), which anchors the protein to the viral membrane (Figure 3A) [40]. The ectodomains (DI, DII, and DIII) form a β-strand-rich surface that is involved in interactions with the host cell membrane and immune system components [40,42–44].

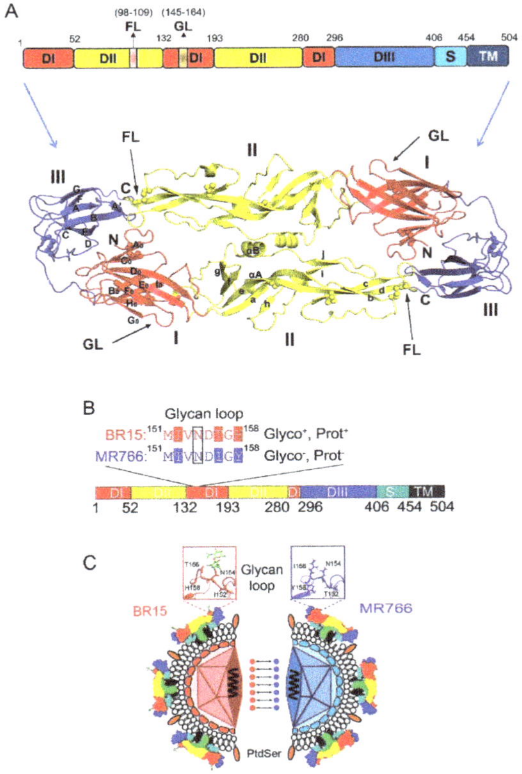

Figure 3. ZIKV E protein and its distinctive structural features between MR766 and BR15 strains. (**A**) Schematic representation of monomer sequence (**top**) and 3-D dimer (**bottom**) of the ZIKV E protein. Structural organization of the E protein monomer, which consists of three N-terminal ectodomains—Domain I (DI), Domain II (DII), and Domain III (DIII)—followed by a C-terminal stem-transmembrane (TM) region that anchors the protein to the viral membrane. Key structural features such as the fusion loop (FL) and glycan loop (GL), along with their respective residue positions, are indicated (**bottom**). Three-dimensional structure of the E protein dimer, highlighting DI, DII, DIII, and the FL. The 3-D E protein structure is adapted from [44]. (**B**) Organizational sequence of the E protein monomer shows differences in glycosylation and protonation status between BR15 and MR766. (**C**) Mirror image comparison of BR15 and MR766 E proteins, highlighting structural differences in the GL. Seven dot connections mark the evolutionarily distinctive a.a. residues within the GL region. The GL shown in the two square boxes illustrates the structural differences between BR15 (red) and MR766 (blue). For indications of different color structures on ZIKV, please see Figure 1A. Abbreviations: Glyco, glycosylation; Prot, protonation state; FL, fusion loop.

Domain I (DI) consists of 132 residues, and it serves as a link to connect DII and DIII, acting as a hinge region that allows flexibility in the protein structure (Figure 3A). DI can be further divided into three segments: the N-terminal region (1–52 a.a.), the central portion (132–193 a.a.), and the C-terminal region (280–296 a.a.) [40,44]. A notable feature within DI is the glycan loop (GL; a.a. 145–164), a flexible structure that can undergo post-translational N-glycosylation at the specific asparagine (N154) residue, potentially influencing viral infectivity [45,46]. Domain II (DII) has a distinctive finger-like structure and is divided into two segments (a.a. 53–131 and a.a. 194–279). It contains a pH-sensitive fusion loop (FL; a.a. 98–109), which is a critical element for the merging of the viral membrane with the host cell membrane during viral entry [44,47]. Domain III (DIII), spanning residues 297–406, plays a crucial role in binding of host cell receptors [41,48–50]. DIII is also a site where several neutralizing antibodies bind, conferring protection against ZIKV infection [51,52]. Moreover, DIII contains an extended CD-loop (a.a. 340–360), which is critical for viral stability and pathogenicity. Disruptive mutations or shortening of the CD-loop destabilize the virus, resulting in significantly reduced viral infectivity and attenuation [53]. This makes the CD-loop an important target for the development of neutralizing antibodies (nAbs) and live attenuated vaccines against ZIKV.

2.2. Post-Translational Modification of ZIKV E Protein

2.2.1. Glycosylation of ZIKV E Protein

The ZIKV E protein can undergo N154 glycosylation, where an N-glycan is attached to the amide nitrogen of the N residue through post-translational modification (Figure 3B) [54,55]. However, the occurrence of glycosylation at the N154 residue of ZIKV E protein depends on several factors. For instance, the presence of serine (S) or threonine (T) at position 156 (154N-X-S/T156, X denotes any residue) is required for glycosylation to take place [20,56]. Glycosylation of the E protein is a critical factor in facilitating viral entry into host cells by interacting with host cell receptors, making it a significant contributor to ZIKV pathogenesis [56]. Additionally, the residues surrounding the N154 glycosylation site, specifically positions at 152, 156, and 158, have been shown to influence conformational changes in the E protein in response to intracellular pH changes [57]. This suggests that the glycosylation status at N154, potentially influenced by these surrounding residues, may also play a role in E-mediated membrane fusion during viral entry. ZIKV E protein exhibits considerable N-glycan heterogeneity across insects such as mosquitos and mammalian cell types, including monocytes, placental, and neural cells, consistent with patterns observed in DENV E protein [45,58]. Routhu et al. showed that the ZIKV E protein displays high heterogeneity N-glycans—from high-mannose (e.g., Man5GlcNAc2) to highly sialylated complex types—depending on the producing cell type. This glycosylation variability is influenced by both viral strain and host cell physiology, potentially impacting ZIKV pathogenesis [45]. Another study confirmed through enzymatic de-glycosylation that the N444 (E154) site in the ZIKV E protein is highly glycosylated and exhibits considerable microheterogeneity from the ZIKV E protein produced in EB66 cells [59]. Additionally, early studies suggest that N154 glycosylation patterns in ZIKV can differ significantly depending on whether the virus is derived from mosquitoes or human cells. For example, a study on the molecular evolution of ZIKV proposed that the loss of the N154 glycosylation site in the E protein may represent an adaptive response to the mosquito vector *Aedes dalzieli* [31]. In another study, genome-wide mapping of functional residues in the ZIKV E protein using a mutant library identified N154 glycosylation as the most divergent determinant of viral replication fitness between mosquito and human cells. This finding highlights the critical role of N154 glycosylation in modulating ZIKV replication efficiency across host cell types [60].

2.2.2. Protonation of ZIKV E Protein

Protein protonation refers to the process of adding a hydrogen ion (proton, H^+) to a specific a.a., such as histidine, which introduces a positive charge to the residue. This can significantly affect the protein's structure and function, depending on the pH of the surrounding environment. The ZIKV E protein contains several potential protonation sites at H158, H249, H288, H323, and H158 [57,61]. Among them, H158 is particularly significant because it resides within the GL of the ectodomain DI (Figure 3B) [57]. During the initial membrane fusion between the virus and the endosomes, the low-pH environment is thought to trigger protonation at H158. This likely induces conformational changes in the GL, disrupting the E protein rafts on the virus surface and leading to the formation of fusion pores that allow the nucleocapsid to be released into the cytosol [57]. However, this mechanism remains to be thoroughly investigated.

2.2.3. Phosphatidylserine on ZIKV E Protein

Phosphatidylserine (PtdSer) is a phospholipid present on the viral envelope of ZIKV and other enveloped viruses, playing a vital role in the viral infection process [62,63]. Findings showed that PtdSer in flaviviruses like DENV normally embedded within the lipid bilayer; PtdSer is interlocked with E proteins to form "raft-like" structures on the viral surface, which help maintain viral stability and function [64]. Under certain conditions, such as specific cell surface receptor signaling, PtdSer can be exposed on the outer surface of the virus [65]. This "flip" of PtdSer from the inner to the outer surface, mediated by scramblases [66], allows the virus to act as a "molecular mimic" of apoptotic cells. This mimicry enables the virus to bind to host cell receptors like AXL through the protein Gas6, facilitating viral entry (Figure 2) [62,67]. Research has shown that disrupting PtdSer on the ZIKV envelope significantly reduces the virus's ability to infect cells [63], underscoring the importance of PtdSer in the viral entry process. However, the precise mechanisms behind this interaction remain underexplored.

2.2.4. Consideration of Epistatic Interaction

Epistatic interactions within ZIKV E protein could also play a crucial role in its function and evolution. Studies have shown that mutations at one site in the E protein can affect the phenotypic consequences of mutations at other sites, influencing traits such as antibody recognition and viral entry [68,69]. As such, mutations in the E protein can disrupt its interaction with prM, leading to a misfolded or non-functional E protein, which then affects its ability to interact with host cell [70]. Another study provided a comprehensive map of how amino acid mutations in the E protein influence viral replication in a permissive cell line and enable escape from two monoclonal antibodies, ZKA64 and ZKA185 [71]. Additional investigation is needed to substantiate the epistatic interactions with ZIKV E protein.

2.3. *Distinctive Features of E Proteins Between Ancestral African and Epidemic ZIKV*

ZIKV can be divided into two major lineages based on their geographic origins and genetic differences [72]. The African lineage is considered the ancestral viruses, while the Asian lineage has been responsible for recent major outbreaks in the Pacific and Americas. Generally, the Asian lineage is considered more concerning due to its potential for causing severe complications like microcephaly when transmitted during pregnancy, while the African lineage typically leads to milder symptoms in humans. Conversely, in animal models, African strains are generally considered more virulent, causing more severe infections and higher mortality rates compared to Asian strains, which tend to be milder [73,74]. Genetic differences exist between these two lineages, particularly in the viral proteins, which contribute to the observed differences in virulence, which can be demonstrated by their evolutionary distance and by comparing their respective viral genomes using phylogenetic tree analysis [75]. Indeed, phylogenetic trees generated by comparing E protein sequences (Figure 4A) also show similar and distinctive genetic distances between African and Asian ZIKV lineages, similar to those phylogenetic trees generated based on the entire viral genomes.

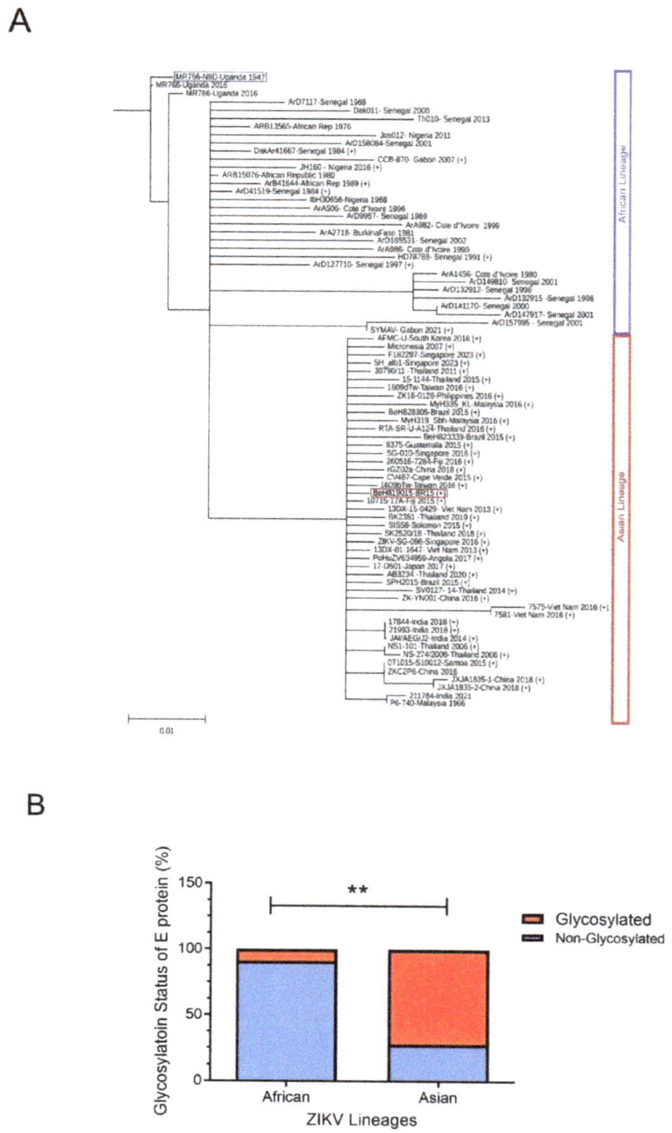

Figure 4. Phylogenetic differences in ZIKV E proteins between African and Asian lineages. (**A**) Phylogenetic tree analysis of ZIKV E protein sequences derived from available database entries. The

tree depicts evolutionary relationships with a specific focus on a.a. residue 156, a key determinant of glycosylation at residue N154. Strains encoding serine (S) or threonine (T) at position 156 are predicted to be glycosylated at N154 [20,56], indicated by a "+" symbol (Gly+), whereas those with isoleucine (I) or other residues at this position are considered non-glycosylated. Two representative strains, BeH8190-BR15 (marked with a red square) and MR766-NIID-Uganda 1947 (marked with a blue square), are highlighted for detailed comparison in the main text. The phylogenetic tree was generated using the neighbor-joining (NJ) and maximum-likelihood (ML) methods, with bootstrap analysis contingent on 1000 replicates based on multiple sequence alignment of ZIKV E proteins using Clustal Omega version 1.2.3. [76]. The unique and complete ZIKV E sequences available in GenBank up to May 2024 were retrieved for this analysis. The resulting trees were visualized using TreeViewer v2.2.0 [77]. The scale bar represents a genetic distance of 0.01, indicating 1% sequence divergence among the aligned E protein sequences. (**B**) Comparison of the relative abundance of glycosylated (**red**) versus non-glycosylated (**blue**) E proteins across lineages. The N-linked glycosylation site at position 154 of the E protein was analyzed among ZIKV strains with unique E protein a.a. sequences, which were obtained from the GenBank virus genome database (http://www.ncbi.nlm.nih.gov/genome/viruses/, accessed on 15 March 2025), Among African lineage viruses, the majority (90.9%; $n = 44$) lack glycosylation at N154, whereas most Asian lineage viruses (72.3%; $n = 36$) are glycosylated at this position. Red box, glycosylated; blue box, non-glycosylated. **. Statistical significance, $p < 0.005$.

Careful examination of the sequence alignments of E proteins reveals some distinctive features that can separate E proteins of ancestral African lineages from those of epidemic Asian ZIKV viruses. To aid in easy comprehension, here we use a pair of ancestral (MR766; GenBank accession#: LC002520) and epidemic (BR15, *aka* BeH819015; GenBank accession#: KU365778) viral sequences to illustrate those differences in E proteins between African and Asian ZIKV lineages (Figure 3B,C). These two viral sequences are chosen as representatives of these two types of viruses because the MR766 ZIKV strain was the first ever reported ZIKV, isolated in 1947 from Africa [5], and BR15, isolated in 2015 from the blood of an epidemic patient in the Pará state [78]. Besides their differences in historic timeline and human disease-causing status [79,80], these two viruses are chosen because, virologically, these two viruses show distinctive contrasts in their ability to bind, infect, replicate, and propagate in neurotropic cells [27,57,81]. For example, testing of chimeric viruses between these two strains by swapping structural proteins suggests that the E protein plays a prominent role in viral binding [27,57,81,82].

Evolutionary analysis with protein sequence alignments of the ancestral MR766 and the epidemic BR15 E protein sequences reveals that BR15 has 15 distinctive a.a. compared to MR766 [81,83]. Of these, seven are evolutionarily distinctive (Figure 3C): three in DI (T152I, I156T, and <u>Y158H</u>) and four in DIII (V317I, <u>I341V</u>, <u>V343A</u>, and D293E), with the three underlined a.a. being human-specific adaptations [81,83]. Notably, all three unique mutations (T152I, I156T, and Y158H) in the DI domain of the epidemic BR15 ZIKV protein are within the GL, which contains both an N-glycosylation site at N154 and a protonation site at H158 (Figure 3C) [84]. In contrast, the ancestral ZIKV MR766 lacks both the N154 glycosylation and the H158 protonation site. The ZIKV E GL extends towards the DII FL and modulates the conformational dynamics of the virion [84], suggesting that the differences in the GL between these two viruses may contribute to differences during viral entry. Most significantly, based on analysis of E protein sequences deposited in GenBank, the majority (90.9%) of epidemic ZIKV E proteins, including BR15, are N154-glycosylated, as they contain an S156 or T156 residue. In contrast, most African isolates (72.3%), including MR766, lack this glycosylation site, and therefore have non-glycosylated E proteins (Figure 4B) [56,85]. The functional importance of E glycosylation and protonation will be further discussed in the following sections.

3. Viral Entry Mediated by ZIKV E Protein

3.1. The Cellular Receptors

The ZIKV E protein mediates viral attachment and fusion with host cell membranes during viral entry. To initiate specific viral attachment and fusion, the ZIKV E protein interacts with multiple cell receptors, including AXL, TYRO3, MER, TIM-1, and DC-SIGN, allowing the virus to efficiently infect a wider range of host cells by providing multiple entry points [86–88]. For example, AXL has been identified as a key cell surface receptor facilitating ZIKV entry across various cell types [89–92]. During viral entry, the ZIKV E protein interacts with the host cell receptor AXL through a bridging molecule called Gas6. Gas6 binds to the viral membrane's PtdSer, effectively connecting the virus to the AXL receptor on the cell surface and facilitating viral entry. Although ProS1 (Protein S) may play a minor role in this interaction, its contribution is generally considered less significant compared to Gas6 in the context of ZIKV infection (Figure 2) [93,94].

By utilizing AXL as a receptor, the ZIKV E protein can efficiently target specific brain cells, particularly neural progenitor cells, which express high levels of AXL, contributing to the virus's neuropathogenesis. For instance, Espino et al. showed the critical role of AXL for ZIKV entry in decidual stromal cells and revealed that blocking the interaction between the ligand–receptor during the initial infection stage significantly reduces virus pathogenesis at the maternal–fetal interface [95]. However, AXL is not the only receptor used during infection in the brain. Studies have shown that even when AXL is inhibited or knocked out, ZIKV can still infect certain brain cells, indicating that alternative receptors may be involved [91,96]. Indeed, research suggests that other TAM receptors such as TYRO3 and MER also play a role in ZIKV infection, albeit to a lesser extent than AXL [97,98]. All of them bind to Gas6 and ProS1 ligands for TAM receptor activation (Figure 2) [93,94]. Furthermore, blocking both AXL and TIM-1 with antibodies can significantly inhibit ZIKV infection, indicating that both receptors may work together to facilitate viral entry [97]. TIM-1, like other TAM family receptors, also binds to PtdSer, facilitating viral adsorption and internalization. However, research indicates that while TIM-1 plays a role, the primary receptor for ZIKV E is AXL, with TIM-1 acting as a secondary or supporting receptor [99]. Therefore, the specific receptors used by the ZIKV E protein depend on the type of cell being infected [100]. An example of receptor-mediated susceptibility is the DC-SIGN receptor, whereby the absence of DC-SIGN on Langerhans cells renders them non-permissive to Zika virus infection, in contrast to DC-SIGN–expressing dendritic cells that support viral entry and replication [101].

3.2. ZIKV E Protein-Mediated Virus Attachment to Host Cells

3.2.1. The Importance of E Protein Structure

The ZIKV E protein is essential for viral attachment to host cells, and any structural modifications can significantly impact viral entry, infectivity, and cell tropism [102]. The DIII of the E protein plays a crucial role in receptor binding [41,48–50], with specific mutations such as A310E and E393K altering cell entry and tropism [102]. Additionally, the CD-loop in DIII is vital for maintaining viral stability, as mutations or shortening of this loop lead to destabilization, reducing infectivity and attenuating the virus [53]. Similarly, a synthesized peptide containing residues 268–273 in DII of the E protein effectively competes with viral attachment and disrupts infection, highlighting the critical role of DII in viral attachment and infectivity [44,70,103].

3.2.2. The Role of N154 Glycosylation

Post-translational modifications of the ZIKV E protein, such as glycosylation, play a crucial role in its interaction with cell surface receptors, influencing viral tropism and infectivity [104]. The ZIKV E protein possesses a well-characterized N-linked glycosylation site at a.a. 154, a highly conserved feature among orthoflaviviruses. This glycan extends from the viral surface and facilitates receptor binding. For instance, glycosylation at N154 enhances ZIKV infection in DC-SIGN-expressing human cells, suggesting its role in receptor interaction [101]. Additionally, viruses retaining this glycosylation site exhibit heightened pathogenicity in mouse models [56,85]. Conversely, viruses lacking glycosylation or those carrying the N154Q glycosylation-deficient mutation display attenuated virulence in mice, leading to lower viral loads in serum and brain tissues [56,85,105]. Interestingly, immunization with the N154Q mutant virus provided protection against wild-type ZIKV (Cambodian strain FSS13025), indicating its ability to elicit a strong antibody response [105].

Studies using pseudoviral particles further confirm the importance of N154 glycosylation, showing a reduction in infectivity when this glycan is absent [106]. Moreover, non-glycosylated viruses replicate inefficiently in mouse brains compared to their glycosylated counterparts, reinforcing the role of N154 glycosylation in ZIKV neurovirulence and invasion [85]. Interestingly, the N-glycans associated with the ZIKV E protein vary in structure and composition depending on viral strain and host cell type [45]. Inhibition studies using mannose-rich competitive inhibitors demonstrated a significant reduction in ZIKV infection in DC-SIGN-expressing cells, suggesting that high-mannose N154 glycans are crucial for interaction with C-type lectin receptors such as DC-SIGN [45]. However, the specific impact of those heterogeneous N-glycans on N154 glycosylated viruses during viral entry, infectivity, and cell tropism has yet been fully investigated. Our own study shows the difference between MR766 and BR15/ICD in viral binding and subsequent viral infection and replication, as well as apoptotic cell death [27,81], suggesting N154 glycosylation of the E protein may play a critical role in viral attachment. However, the molecular action underlying how N154 glycosylation affects the viral binding remains unclear.

The E glycosylation appears to be critical for ZIKV infection of mammalian and mosquito cells, because a glycosylation mutant N154Q diminished oral infectivity by *Ae. aegypti* vector and showed reduced viremia and diminished mortality in mouse models [105]. While non-glycosylated ZIKV displayed reduced neuroinvasion when introduced subcutaneously, it replicated efficiently following intracranial inoculation, implying a role for E glycosylation in BBB penetration [85]. Furthermore, ZIKV viral particles lacking the E protein glycan were still able to infect Raji cells expressing the lectin DC-SIGN receptor, indicating the prM glycan of partially mature particles can facilitate the viral entry [84]. The E protein, specifically its extended CD-loop, may confer viral stability, cell cycle-dependent viral replication, and in vivo pathogenesis, as shortening the CD-loop compromise the virus, and Δ346 mutation in this loop disrupts thermal stability of the virus [53].

3.2.3. The Role of PtdSer

PtdSer, a phospholipid primarily found on the inner leaflet of cell membranes on the viral envelope, has been implicated in facilitating the binding of enveloped viruses [63]. Since the phospholipid PtdSer is interlocked with E proteins to form "raft-like" structures on the virus surface and serves as a receptor-binding ligand, it may also play a role in viral attachment and viral entry [62,64,107]. Limited studies have shown that PtdSer indeed support viral entry and ZIKV infection in human and mosquito cells as well as in a mouse model [63]. Furthermore, a PtdSer-specific inhibitor could reduce the viral load in serum and spleen in the mouse model [63]. One interesting feature of PtdSer, observed in DENV, is its exposure to the outer viral surface upon an increase in temperature. For example, at a low temperature of 4–28 °C, orthoflavivirus virions display a "closed smooth surface morphology" where PtdSer are hidden behind the viral envelope. Thus, they are less efficient in attaching to host cells. In contrast, when the temperature is shifted to physiologically relevant human body temperature of 37 °C, virions show a "bumpy morphology" where PtdSer are extended out from the viral envelope. As a result, they bind to host cells with high efficiency [108]. As glycosylated and non-glycosylated ZIKV binds to host cells differently, besides N154 E glycosylation of E protein, another possibility to explain the observed differences in viral entry, infectivity, and cell tropism could be the phospholipid variation on the virus surface. However, no research has yet been done in this regard.

3.3. The Role of ZIKV E Protein in Endocytosis and Viral Fusion

Following E protein-mediated viral attachment to host cells, ZIKV enters host cells primarily through clathrin-dependent endocytosis (CDE) (Figure 2) [1,109], although clathrin-independent endocytosis (CIE) might also occur [109]. After the virus is internalized *via* clathrin-coated pits, it is enclosed in an endocytic vesicle and transported to early Rab5+ endosomes (pH 6.8–6.3), where it initiates infection. As the endosome matures into a late Rab7+ compartment and the pH declines to approximately 6.2–5.0, the E protein undergoes a pH-dependent conformational shift. This exposes its FL, facilitating the fusion of the viral envelope with the endosomal membrane. Consequently, the viral RNA genome is released into the host cell cytoplasm within 10 to 15 min of infection [110,111].

A comparative study of viral entry between the African and Asian ZIKV lineages reveals that the clathrin-dependent entry mechanism is highly conserved across both lineages [111]. ZIKV predominantly fuses with late endosomes, likely due to the low pH in these endosomes inducing a conformational change in the ZIKV E protein, which facilitates its interaction with the host cell membrane and subsequent fusion [111]. The E-152/156/158 residues of the ZIKV E protein are believed to be crucial in the later stages of viral entry, particularly in facilitating the conformational changes required for membrane fusion with the host cell [57]. This may be due to their location within the GL of the E protein, where notable differences in N154 glycosylation and H158 protonation status exist between the African and Asian lineages (Figure 3B,C).

4. ZIKV E Protein-Mediated Immune Response and Evasion

The protective immune response serves as a first-line defense of the cellular immune system against viral infections, typically initiated by the release of cytokines and chemokines. As the major surface protein of ZIKV, the E protein functions as the principal viral antigen, eliciting host innate and adaptive immune responses [44]. Notably, N154 glycosylation of the E protein plays a critical role in modulating interactions with the host immune system.

A study investigating ZIKV infection in human-induced pluripotent stem cell (hiPSC)-derived macrophages and microglia found that an epidemic Brazilian ZIKV strain (GZ01/2016; KU820898) carrying a glycosylated N154 E protein triggered a stronger immune response than the non-glycosylated MR766 strain. The glycosylated strain induced higher levels of immune mediators, including AIF1, IL-6, IL-1β, IL-10, G-CSF, and CCR5 [87]. Furthermore, the expression of AXL and TYRO3, both TAM receptors, was significantly elevated in cells infected with the epidemic strain compared to MR766, suggesting that N154 glycosylation enhances viral attachment and immune activation [87].

Interestingly, a contrasting effect of E protein glycosylation was observed in primary mouse dendritic cells, where N154 glycosylation led to a weaker immune response. The glycosylated virus elicited lower levels of IFN-α, IFN-β, IL-1β, and IL-6 compared to the non-glycosylated N154Q mutant [105]. The same study also demonstrated that E protein glycosylation negatively regulated viral attachment, virion assembly, and progeny infectivity in the mosquito C6/36 cell line [105].

These discrepancies may arise from differences in host cell receptor expression and immune responses that influence viral replication. They also suggest that the impact of E protein glycosylation is host- or context-dependent. For example, glycosylation may influence viral interactions with different receptors or modulate distinct cellular pathways, leading to variable outcomes depending on the host species or cell type involved. Another possible explanation is that the role of E protein glycosylation depends on the specific cell surface receptors engaged during infection. In the human macrophage study, TAM receptors were prominently upregulated [87], whereas in mouse dendritic cells, ZIKV primarily utilized DC-SIGN as the entry receptor [105].

Supporting this notion, our recent data show that the epidemic BR15 strain, which carries a glycosylated N154 E protein, induced higher immune responses than the non-glycosylated MR766 strain in the neuroblastoma SH-SY5Y cell line. Similar to the findings of Mesci et al., BR15 triggered elevated levels of IL-6, IL-1β, IL-10, and CCR5 [87]. Additionally, we observed activation of other proinflammatory cytokines, including IL-8, CCL2 (MCP-1), and CCL5 (RANTES) (our unpublished data). However, in contrast to Mesci's findings, our results indicate that the N154-glycosylated BR15 virus exhibited reduced viral binding, replication, and apoptotic cell death compared to the non-glycosylated MR766 strain. Although other viral factors may contribute to these differences, receptor expression variations between cell types could potentially explain these discrepancies. Unlike hiPSC-derived macrophages and microglia, which express AXL and TYRO3, SH-SY5Y cells predominantly express AXL and MER [112]. These differences in receptor engagement may influence the extent of immune activation and viral replication, highlighting the complexity of E protein glycosylation in ZIKV pathogenesis.

One potential argument against using the MR766 strain for E glycosylation studies is that this strain has been passaged too many times in vitro, potentially leading to non-representative results [79]. However, the only difference between the glycosylated and non-glycosylated viruses in the Fontes-Garfias' study [105] was the N154Q mutation in the E protein, underscoring the importance of E glycosylation in these findings. Additionally, the role of N154-glycosylation in eliciting host immune responses is supported by numerous clinical reports. Similar proinflammatory cytokines like IL-1β, IL-6, IL-10, and G-CSF were elevated in the blood, brain, cerebrospinal fluid, and amniotic fluid of ZIKV-positive pregnant women [113–116]. Lastly, animals inoculated with recombinant ZIKV E protein alone also elicit various protective and proinflammatory cytokines, including IFNγ, TNFα, IL-4, IL-6, and IL-12 [117–119].

Besides triggering host cellular inflammatory immune responses as described above, ZIKV also activates the production of Type I interferon (IFN) as part of the host antiviral response [1,120]. This antiviral IFN response is facilitated by mechanisms involving the activation of Toll-like receptors (TLRs), particularly TLR3 and TLR7/8, as well as the RIG-I-like receptor (RIG-1) [1,120]. TLR3- and RIG-1-mediated Type I IFN production, followed by activation of the JAK/STAT pathway, help the host cell mount resistance against ZIKV infection [121].

During the initial stages of infection, ZIKV interacts with host cell surface receptors such as AXL, which not only promotes viral entry but also suppresses host Type I IFN response. This occurs through the interaction of ZIKV E protein with Gas6, triggering the AXL signaling pathway and allowing immune evasion by suppressing the antiviral response [91,122]. The role of ZIKV E protein in immune evasion has been demonstrated through experiments using the human glial cell line SNB19, where Atranorin, a secondary metabolite of lichens, inhibited ZIKV infection by attenuating the activation of the IFN signaling pathway through direct targeting of the E protein [123].

In addition to evading IFN responses, ZIKV infection activates the host complement system, a key component of both the innate and adaptive immune systems. The complement system functions to defend against viral infections by forming the membrane attack complex (MAC), which creates pores in the viral membrane, leading to virolysis or complement-mediated lysis [116,124]. ZIKV has developed mechanisms to counteract the complement system. One study found that ZIKV E protein interferes with the formation of the MAC by binding to terminal complement proteins like C9, reducing polymerization and thereby preventing effective complement-mediated virolysis [125]. This direct interaction with the complement system allows ZIKV to evade destruction by the host immune response.

N154 glycosylation of the E protein may also enhance viral infection by aiding in the evasion of the host's innate antiviral response. For instance, the glycosylated E protein inhibits the reactive oxygen species (ROS) pathway, which is essential for ZIKV infection in *Aedes* mosquitoes. Notably, mutating the glycosylation-promoting residue T156 to either A or I, thereby removing N154 glycosylation, prevents mosquito infection [126]. This phenotype can be rescued by inhibiting the ROS pathway, suggesting that the glycosylated E protein facilitates viral evasion of the mosquito's innate immune defenses to enhance infection. Supporting this notion, *Aedes* mosquitoes infected with the N154-glycosylated Cambodian ZIKV strain FSS13025 exhibited significantly higher infection rates compared to those infected with the non-glycosylated N154Q mutant of the same virus [105]. Similarly, in an immune-compromised IFNAR$^{-/-}$ A129 mouse model, E glycosylation substantially increased mortality, accompanied by high viremia. In contrast, the N154Q mutation markedly attenuated ZIKV, as evidenced by reduced viremia, minimal weight loss, and complete survival [105].

5. Neutralizing Antibodies, Antibody-Dependent Enhancement, and Vaccine Development

5.1. Neutralizing Antibodies (nAbs)

The nAbs are crucial components of the host immune response against ZIKV infection. The E protein, being the primary antigen on the surface of the virus, serves as a major target for nAbs. These antibodies typically prevent infection either by blocking the virus from binding to its host cell receptors or by inhibiting the structural changes required for the E protein to mediate fusion during viral entry via the endocytic pathway [127].

As described earlier, the ZIKV E protein consists of three structural domains: DI, DII, and DIII (Figure 3A). Domain I serves as a connector between DII and DIII, acting as a flexible hinge that enables the conformational changes necessary for viral entry into host cells [128]. While all three domains can be targeted by nAbs, DIII is particularly important because it is responsible for host cell receptor binding, making it the primary target of nAbs against the ZIKV [52,129]. Due to this critical role in neutralization, DIII is often a key focus in ZIKV vaccine development [52].

Numerous DIII-specific nAbs have been isolated from both convalescent patients and immunized mice (Table 1). Three distinct epitopes in the DIII domain have been identified: the lateral ridge (LR), the C–C′ loop, and the ABDE sheet region, which is a specific region that is highly conserved among different ZIKV sequences [130]. The LR is a prominent structural feature formed by loops connecting β-strands, while the C–C′ loop is a flexible region that plays a vital role in receptor binding and shows variability across ZIKV strains [131]. The ABDE sheet is a β-sheet region that acts as an important epitope for nAbs [70].

In early studies using X-ray crystallography and competition binding analyses, six mouse monoclonal antibodies (mAbs) against ZIKV were analyzed. Among these, ZV-48 and ZV-64 bind to the C–C′ loop, ZV-2 binds to the ABDE sheet, and ZV-54 and ZV-67 target the LR of the DIII domain. ZV-48, ZV-54, ZV-64, and ZV-67 were specific to ZIKV and neutralized all ZIKV lineages [132]. However, only ZV-67 neutralized all ZIKV strains in vitro and provided protection in lethal challenge models in vivo [132]. Other mAbs, such as ZIKV-116, also target the LR and prevent ZIKV infection by blocking viral fusion with host membranes [131]. However, ZIKV-116 cross-neutralizes certain strains of DENV1, with higher neutralizing activity observed against Asian ZIKV strains compared to African strains [131]. Additional mAbs, such as Z004, Z006, and Z021, are part of a set of recurrent antibodies that neutralize both ZIKV and DENV [133,134]. In a separate study, five DIII-specific nAbs (HA-12, 1C-11, 2F-8, 1D-9, and 1D-11) were isolated from ZIKV-infected patients, all of which neutralized the Asian strain of ZIKV. Among these, HA-12 and 2F-8 also neutralized American strains, i.e., ZIKV of Asian lineage found in the America. The most potent antibody, 2F-8, conferred complete protection in mice when administered in a single dose, either before or after exposure to a lethal ZIKV challenge [135].

Besides targeting DIII, the junction between DI and DIII is another important site for nAbs. This region forms a quaternary epitope accessible only when viral proteins are in their native conformation. Antibodies targeting this junction can effectively block viral entry and neutralize infection by disrupting the virus's ability to fuse with host cells, potentially offering high specificity to ZIKV while minimizing cross-reactivity with related viruses [129]. For example, the nAb B11F, recovered from a convalescent patient, targets a novel epitope that spans both DI and DIII [136]. Additionally, four other nAbs (ZKA190, MZ1, MZ4, and MZ24) target the DI/DIII linker region of the ZIKV E protein. ZKA190 demonstrated high efficacy in vivo, preventing morbidity and mortality in ZIKV-infected mice, while MZ1, MZ4, and MZ24 exhibited cross-neutralizing activity against both ZIKV and DENV [137,138].

In addition to DI and DIII, quaternary epitopes in the DII domain of the ZIKV E protein are also targeted by nAbs [139]. The nAb ZKA185, for instance, targets DII and neutralizes both African and Asian ZIKV lineages [140].

The FL, located between the DII and DIII domains, is another important target for nAbs due to its role in viral fusion with host membranes. Since this region is conserved across orthoflaviviruses and overlaps with the prM binding site in immature particles, it presents a target for broad-spectrum nAbs [141]. One such antibody, 2A10G6, provided broad protection against ZIKV, DENV1-4, and WNV in mouse models and in vitro studies. Human-derived mAbs such as AZ1p and AZ6m, which target the ZIKV E protein FL, also demonstrated broad binding capacity and neutralization of ZIKV, YFV, and DENV [142].

The GL within DI of the ZIKV E protein is also a target for nAbs due to its role in modulating E protein conformational changes during viral fusion [57,84]. A recombinant ZIKV protein (E80) covering 80% of the N-terminal ectodomain-induced potent nAb responses protected mice from lethal challenges of the ZIKV infection [51]. Subsequently, five mAbs (3E8, 5F8, 5G3, 8A2, and 9C3) were isolated from E80-immunized mice, showing specific neutralization of Asian-lineage ZIKV strains but not African strains [143]. Among these, 5F8 conferred complete protection in a mouse model by inhibiting viral entry during the early post-attachment stage. The epitope recognized by 5F8 is highly conserved in ZIKV but varies between ZIKV and other orthoflaviviruses, possibly representing a new class of ZIKV-specific nAbs targeting the GL [143]. However, other studies have shown that certain mAbs targeting the GL, such as A11 and A42, exhibit broader activity against ZIKV and other orthoflaviviruses [144]. These findings underscore the importance of the GL as a key antigenic determinant [43,82].

Another study further emphasized the role of the GL in nAb-mediated ZIKV neutralization [82]. As previously described, the GL contains three evolutionarily distinctive residues, 152, 156, and 158, within the ZIKV E protein (E-152/156/158), which in most cases can distinguish the African lineage from the Asian lineage (Figure 3B). To investigate the role of N154 glycosylation and H158 protonation in nAb-mediated ZIKV neutralization, researchers replaced the E-152/156/158 residues of the BR15 strain with those from MR766 (I152T, T156I, and H158Y), resulting in a non-glycosylated virus in the BR15 backbone [82]. When this chimeric ZIKV was inoculated into adult BALB/c mice, it rapidly induced nAb production. Interestingly, while the nAbs effectively neutralized the MR766 strain, they failed to neutralize the BR15 clone. This suggests that the E-152/156/158 residues in the GL, or the N154 glycosylation, may influence nAb epitope accessibility, and that differences in N154 glycosylation between MR766 and BR15 may lead to variations in neutralization [82].

In a follow-up study [43], the role of the GL-residing E-152/156/158 residues as antigenic determinants was further tested. An oligopeptide consisting of a 20-mer (a.a. 145–164) from the GL, which carried the non-glycosylated MR766-derived residues E-T152/I156/Y158, was used to assess immunoreactivity against anti-GL Abs from the serum of the same chimeric ZIKV-challenged mice described earlier [82]. Results showed that the 20-mer with MR766's E-T152/I156/Y158 residues reacted with anti-GL Abs from the BR15-E-T152/I156/Y158-challenged mice. However, the 20-mer carrying BR15-derived E-I152/T156/H158 did not react with the same serum, suggesting that anti-GL Abs induced by the chimeric virus were specific to MR766. Notably, changing residues T152 to I152 and I156 to T156 did not alter the immunoreactivity of the anti-GL sera; whereas, the substitution of Y158 with H158 (Y158H) abolished the reactivity to the anti-GL sera. Structural analysis of the E-Domain I, including both the wild-type 20-mer and mutant E-T152I and E-I156T 20-mers, suggested that residues E-152 and E-156 influence the conformation of the GL. This alteration in GL structure could affect the antigenic reactivity of Abs targeting this region. While these studies did not specifically evaluate the roles of N154 glycosylation or H158 protonation, the findings suggest that the E-152/156/158 residues are likely key antigenic determinants within the ZIKV GL region [43].

The use of nAbs that target multiple domains of the ZIKV E protein could maximize therapeutic potential and prevent viral escape [133,140]. For instance, human-derived nAbs Z23 and Z3L1, which target tertiary epitopes in DI, DII, and DIII, exhibit potent ZIKV-specific neutralization and offer post-exposure protection in mice [145]. Immunization with the E protein or its subdomains (DI/DII and DIII) has also been shown to induce high titers of E-specific antibodies that neutralize ZIKV [119].

Finally, The E dimer epitope (EDE) of the ZIKV E protein, a unique feature formed by the interaction of two E protein dimers on the viral surface, is another significant target for nAbs due to its high specificity and ability to block viral entry [146]. EDE-targeting antibodies, such as EDE1-C8, EDE1-C10, and EDE1-B10, which were initially raised against DENV, also exhibit cross-neutralizing activity against ZIKV [139,147]. In mice, a single dose of EDE1-B10 administered three days post-infection, prevented death and reduced ZIKV levels in the brain and testes [148]. Similarly, the human-derived nAb ZIKV-117, which targets EDE, significantly reduced tissue damage, placental and fetal infection, as well as mortality in pregnant and non-pregnant mice [149,150].

Table 1. List of nAbs against ZIKV envelope protein.

Neutralizing Antibody	Target Site	Specificity	Reference
ZV-2, ZV-48, ZV-54, ZV-64, ZV-67	DIII or DIII LR	ZIKV	[132,149]
ZIKV-116	DIII	ZIKV	[131,149]
Z004, Z006, Z021	DIII LR	ZIKV and DENV1	[134]
HA-12, 1C-11, 2F-8, 1D-9, 1D-11	DIII	ZIKV Asian lineage	[135]

Table 1. *Cont.*

Neutralizing Antibody	Target Site	Specificity	Reference
B11F	DI and DIII	ZIKV	[136]
ZKA190	DI-DIII linker, DIII LR	ZIKV	[137]
MZ1/MZ4/MZ24	DI/DIII linker region	ZIKV, DENV1-4	[138]
ZKA185	DII	ZIKV	[140]
2A10G6	DII (FL)	ZIKV, DENV1-4 and WNV	[44,141]
AZ1p, AZ6m	DII(FL)	ZIKV, YFV, and DENV	[142]
3E8, 5F8, 5G3, 8A2, 9C3	Linear Epitope on GL	ZIKV Asian lineage	[143]
A11, A42	GL	ZIKV	[144]
Z3L1, Z23, Z20	DI, DII, or DIII	ZIKV	[145]
rhMZ—Group A, B, C, D	EDE	ZIKV	[146]
EDE1-B10, EDE1-C8, EDE1-C10	EDE	ZIKV and DENV 1-4	[69,139,147]
ZIKV-117	EDE	ZIKV	[149,150]

5.2. Antibody-Dependent Enhancement (ADE) and Vaccine Development

ADE presents a significant challenge in the development of effective therapies against ZIKV infection. ADE occurs when non-neutralizing or sub-neutralizing levels of nAbs bind to ZIKV but fail to block the virus from infecting host cells. Instead, the antibody–virus complex facilitates viral entry into cells via Fc receptor-mediated internalization, leading to increased rates of infection and potentially more severe disease outcomes [151,152]. This phenomenon is particularly concerning in regions where both ZIKV and DENV co-circulate, as cross-reactive antibodies generated during primary DENV infection may enhance ZIKV infection upon secondary exposure, exacerbating disease severity [153,154]. Therefore, understanding ADE is essential for developing safe and effective vaccines against ZIKV infection [155].

Note that the occurrence of ADE in ZIKV infection remains an area of active investigation. Only limited clinical evidence suggests that ADE may occur, particularly in the context of pre-existing immunity to DENV. For instance, pre-existing DENV antibodies have been shown to influence the antibody response during primary ZIKV infection, leading to enhancement of ZIKV infection in vitro and in vivo [156,157]. An epidemiological study also found that convalescent sera from dengue patients exhibited ADE for ZIKV [158]. Conversely, a fatal case of DENV following prior ZIKV infection has been interpreted as a clinical example of ADE [159]. In addition to these clinical observations, several experimental studies have demonstrated that DENV-immune sera or monoclonal antibodies can enhance ZIKV infection through Fc receptor-mediated uptake in cell cultures and animal models [160,161]. Although the extent of ADE in natural ZIKV infection remains uncertain, these findings collectively support the possibility that ADE may occur in ZIKV under certain immunological conditions, especially in flavivirus-endemic regions.

As ZIKV E protein is a primary target for nAbs, many studies have focused on the immune responses elicited by this protein, including cross-reactivity with antibodies against other orthoflaviviruses such as DENV [44,162]. ZIKV and DENV are closely related viruses within the flavivirus family, sharing similar modes of transmission, clinical symptoms, and mosquito vectors, particularly *Aedes aegypti* [163]. DENV is the most prevalent orthoflavivirus in South America and represents the leading cause of orthoflavivirus-related diseases in Brazil [164]. However, the rising global temperatures have contributed to the spread of DENV, increasing its threat to public health in other regions, including the United States [165]. For instance, the first case of fetal DENV infection was reported in the U.S. in 2021, where the child had prior exposure to ZIKV, an outcome attributed to ADE [159]. Due to the high degree of similarity between ZIKV and DENV, about 55% a.a. identity in the E protein, ADE frequently arises from cross-reactive nAbs generated during a primary DENV infection, leading to an enhanced risk of secondary infection with ZIKV [166,167].

One approach to mitigating the risk of ADE in vaccine development is to identify ZIKV-specific nAbs that do not cross-react with other orthoflaviviruses, particularly DENV. For example, to minimize cross-reactivity between ZIKV and DENV, one study isolated four groups of EDE-specific nAbs from ZIKV-infected rhesus macaques. These antibodies not only protected against ZIKV replication in mice but were also identified in convalescent humans following ZIKV infection [146]. Moreover, a study showed that plasma from individuals previously infected with DENV could enhance ZIKV infection in human myeloid cells via ADE [168]. The mathematical modeling study also revealed that dengue vaccination, especially with imperfect pre-vaccination screening, can enhance ZIKV transmission through ADE, underscoring the need for cautious implementation in regions where both viruses co-circulate [169].

In another study, SIgN-3C, a human-derived antibody initially raised against DENV1-4, also neutralized ZIKV infection in an IFNAR$^{-/-}$ mouse model [170,171]. However, a variant of this antibody, SIgN-3C-LALA, did not induce ADE in vitro and provided similar protection in vivo [171]. In pregnant ZIKV-infected IFNAR$^{-/-}$ mice, treatment with SIgN-3C or SIgN-3C-LALA significantly reduced viral loads in fetal organs and placenta, mitigating virus-induced fetal growth retardation [171].

A different study has identified six ZIKV-specific nAbs that target key epitopes on the ZIKV E protein, including the C-C′ loop and the LR of the DIII domain (e.g., ZV-48, ZV-54, and ZV-67) [132]. These nAbs offer potential for targeted vaccine development by focusing on epitopes unique to ZIKV, thereby reducing the risk of cross-reactivity with DENV and minimizing the potential for ADE.

A second strategy involves modifying the Fc region of ZIKV-specific antibodies to prevent ADE while maintaining their neutralizing ability. In one study, researchers used a prior developed ZIKV-specific nAb ZV-54 targeting the LR of the DIII domain and then modified its Fc glycosylation profile to prevent interaction with Fcγ receptors on immune cells. The ZV-54 variants retained their neutralizing potency against ZIKV without triggering ADE in DENV infection, demonstrating a promising dual approach that preserves efficacy while eliminating ADE risk [172]. This approach showcases the potential of combining ZIKV-specific nAbs with targeted modifications to enhance vaccine safety.

Target-specific mutations in the ZIKV E protein may also improve the specificity of nAbs and reduce ADE risk. Several novel mutations in or near the FL of DII or DIII have been reported to decrease the production of cross-reactive antibodies [173]. These mutations may alter the antigenic structure of the E protein, making it more challenging for cross-reactive antibodies generated during DENV infection to enhance ZIKV infection. By improving the specificity of nAbs, these mutations can enhance the safety of vaccines by reducing the likelihood of ADE.

In addition to modifying the viral antigens, engineering nAbs to prevent their interaction with Fcγ receptors on immune cells offers another potential solution to ADE. This can be achieved through mutations in the Fc region of the nAbs, which block Fcγ receptor engagement and thereby prevent antibody-dependent viral entry into host cells [173]. Other techniques to limit ADE include blocking Fc receptor engagement, removing the antibody heavy chain, or deleting the N-linked sugar on IgG molecules, all of which prevent ADE while maintaining the neutralizing capacity of the antibodies [174].

In conclusion, strategies to reduce the risk of ADE in ZIKV vaccine development include identifying ZIKV-specific nAbs, modifying the Fc region of antibodies to avoid Fcγ receptor interactions, and introducing mutations in the ZIKV E protein to improve nAb specificity. These approaches are critical for creating safe and effective vaccines and therapies against ZIKV, while minimizing the potential for cross-reactivity with other orthoflaviviruses and reducing the risk of ADE.

6. Anti-ZIKV E Protein Inhibitors

As discussed in the previous section, vaccine development for ZIKV faces significant challenges, including cross-reactivity of nAbs, ADE, and antigenic diversity among different ZIKV lineages and other orthoflaviviruses [168]. Given these limitations, the discovery of inhibitors or host cellular restriction factors targeting the ZIKV E protein presents a promising alternative strategy to block viral entry and limit infection. As ZIKV E protein plays a critical role in mediating viral entry by facilitating multiple steps of virus–host interactions [1,86,88]. Depending on their mechanism of action, ZIKV E protein inhibitors can be categorized into four groups: (1) direct E protein inhibitors: molecules that interact with the E protein to block its function, (2) viral attachment inhibitors: agents that prevent ZIKV from binding to host cells by disrupting E protein interactions with cell surface receptors, (3) viral fusion inhibitors: compounds that block viral fusion with endosomes, thereby preventing genome release, and (4) host cellular restriction factors: cellular factors that interfere with ZIKV E protein-mediated viral entry. ZIKV E protein inhibitors can take various forms, including small molecules, natural compounds, proteins, and peptide-based inhibitors. Table 2 provides a comprehensive list of known inhibitors that target the ZIKV E protein or its associated viral entry processes. A brief summary of their mechanisms of action is outlined below.

6.1. Direct E Protein Inhibitors

Most of the direct E protein inhibitors reported so far are natural compounds with few exceptions. A small molecule F1065-0358 was found to inhibit ZIKV infection by binding to a region between DI and DIII of the ZIKV E protein, preventing its trimerization, a crucial step in the fusion process required for viral entry [175]. A natural compound, gossypol, a phenol derived from the cotton plant, exhibits potent antiviral activity against both African and Asian ZIKV strains. It neutralizes ZIKV infection by targeting DIII of the E protein. When combined with the anti-cancer drug bortezomib, it achieves a synergistic enhancement of antiviral activity [176,177]. Epigallocatechin gallate (EGCG), a natural compound found in green tea, binds to DI and DII of the ZIKV E protein, disrupting the conformational changes necessary for viral entry [178]. Pentagalloylglucose (PGG), a polyphenolic compound from the Chinese herb, inhibits ZIKV entry by interacting with charged residues on the viral envelope, such as glycosylated viral E proteins, thereby competing with natural virus–receptor interactions essential for viral binding [179]. Polysaccharides like Pentagalloylglucose, parishin, and stevioside inhibits the viral entry by binding the Zika virus envelope protein [180].

Other natural compounds include Apigenin, a flavonoid from the flavone glycoside class, which exhibits high affinity for the ZIKV E protein DIII and significantly reduces viral titers in Vero cells [181]. Similarly, baicalin, a glucuronide derivative of baicalein found in *Scutellaria baicalensis*, has demonstrated strong antiviral activity against ZIKV entry with high binding affinity to the E protein and low cytotoxicity [182]. Atranorin, a lichen-derived secondary metabolite, effectively blocks ZIKV entry and reduces infectivity in vitro by directly targeting the E protein, with docking studies suggesting optimal binding between DI and DIII. Additionally, it inhibits ZIKV infection in human glial cells by attenuating

IFN signaling, further highlighting its potential as an antiviral agent [123]. Palmatine, a protoberberine alkaloid compound that inhibits ZIKV infection by interacting with the E protein, has been shown to destabilize ZIKV. It inhibited ZIKV binding and entry by 95% and 69%, respectively, likely through interaction with the ZIKV E protein, consistent with molecular docking analysis [183]. Lastly, harringtonine, a natural alkaloid from the Cephalotaxus genus, demonstrates nanomolar-level efficacy in inhibiting multiple stages of ZIKV infection, including stability, binding, entry, replication, and release, through direct interaction with the E protein [184].

6.2. Viral Attachment Inhibitors

Several attachment inhibitors, Cabozantinib, R428, TP-0903, and BMS-777607, significantly impaired ZIKV infection by inhibiting the AXL receptor in a human cerebral microvascular endothelial hCMEC/D3 cell line and the human umbilical vein endothelial cells (hUVECs) [185], which may serve as potential antiviral therapeutics for suppressing ZIKV infection [185,186]. Another small molecule attachment inhibitor ZINC33683341, identified through in silico modeling, has been shown to inhibit ZIKV entry by binding to primary receptors, with its antiviral activity confirmed in in vitro studies [187]. Curcumin, a component of turmeric, inhibits ZIKV attachment and entry by disrupting viral E protein function and blocking E-cell receptor interactions [188,189].

6.3. Viral Fusion Inhibitors

A number of inhibitors have been identified that target the viral fusion step of the ZIKV E protein during viral entry. Through a high-throughput screening method, known as ALPHAscreen, which is a competitive amplified luminescent proximity homogeneous assay, has been used to discover small molecule inhibitors of the ZIKV E protein. Seven lead compounds were identified to disrupt E protein-mediated membrane fusion during viral entry of ZIKV and DENV infections, thereby blocking the viral infection [190]. Moreover, the recombinant venom peptide rEv37, derived from the scorpion *Euscorpiops validus* and produced in a prokaryotic system, has shown to prevent the low pH-dependent fusion, thus restricting the late step of the viral entry into the host cell during ZIKV and DENV2 infection [191]. Atovaquone, a synthetic compound with antifungal antiparasitic properties [192,193] has been shown to suppress ZIKV and DENV1-4 infection in both mammalian and mosquito-derived cells [194]. Its antiviral activity is attributed to the inhibition of E protein-mediated membrane fusion during viral entry [194]. The synthetic peptide Z2 disrupts ZIKV virion integrity, inhibiting infection in vitro. Additionally, it crosses the placental barrier and prevents vertical transmission in C57BL/6 pregnant mice, possibly by inducing pore formation in the viral membrane through disruption of E protein conformational changes [195,196]. Another fusion inhibitor, P5, a peptide derived from the JEV E protein stem, effectively blocked ZIKV infection and conferred in vivo protection in AG6 mice by altering E protein conformation under low pH conditions [197].

6.4. Host Restriction Factors

Several host cellular factors restrict ZIKV infection, including LAMR1 (Laminin receptor 1), Hpa (heparanase), Viperin, and USP38 (Ubiquitin-specific peptidase 38), which target different stages of the viral life cycle, such as entry, replication, and protein stability. Specifically, LAMR1 suppresses ZIKV infection by interacting with the E protein and reducing its ubiquitination [198]. Hpa, a multifaceted protein and an endo-β-D-glucuronidase, degrades heparan sulfate and functions as a host restriction factor to attenuate ZIKV infection by destabilizing the viral E protein [199]. Viperin is an interferon-inducible protein, also known as virus-inhibitory protein, restricting the replication of a wide range of viruses including controlling the ZIKV infection [200]. Finally, USP38, a ubiquitin-specific peptidase, has been identified as a host restriction factor in resisting ZIKV infection by removing the ubiquitination of the viral E protein, which is critical for ZIKV infection and transmission [201].

Table 2. Anti-ZIKV E inhibitors.

Compound Name	Effect	IC_{50}/EC_{50} (μM)	CC_{50} (μM)	Cell /Model	Reference
Direct E Protein Inhibitors					
F1065-0358	Bind to the DI and DIII regions and interfere with the E protein trimerization during	14	200	Vero	[175]
Gossypol	Bind to DIII	3.75 ± 0.01	14.17 ± 0.74	Vero E6	[177]
EGCG	Bind to DI and DII	na	na	Vero E6	[178]
PGG	Interacts with charged residues of glycosylated E protein	4.1	114	Vero B4	[179]
Polysaccharides (PGG, parishin and stevioside)	Bind to E protein	na	na	Docking analysis	[180]
Apigenin	Bind to DIII	>100	na	Vero	[181]
Baicalin	Bind to E protein	14	na	Vero	[181,182]
Atranorin	Bind to DI and DIII	11.9	>50	SNB-19	[123]
Palmatine	Interact with E protein	na	na	Vero	[183]
Harringtonine	Bind to E Protein	0.287	>10	Vero	[184]
Viral Attachment Inhibitors					
Cabozantinib (R428, TP-0903, and BMS-777607)	Inhibit AXL receptor	na	na	hCMEC/D3/HUVECs	[185]
ZINC33683341	Bind to primary receptors	na	na	Vero	[187]
Curcumin	Viral attachment and entry by abrogating the function of viral envelope proteins	1.9	11.6	Vero	[188,189]
Viral Fusion Inhibitors					
Seven compounds	Prevent E-mediated membrane fusion	0.9–19.3	5.2–>100	Vero	[190]
Ev37	Prevent viral membrane–endosomal membrane at low pH	na	116.3	Huh-7	[191]
Atovaquone	Block E-mediated membrane fusion	2.1	na	Vero/MDCK/C6/36	[194]
Peptide Z2	Disrupt E conformational changes	1.75	na	C57BL/6BHK21	[195,196]
P5	Change E protein conformation at low pH	3.27	na	Vero/AG6	[197]
Host Restriction Factors					
LAMR1	Attenuate E protein ubiquitination	na	na	HeLa/HEK293T	[198]
Hpa	Attenuates ZIKV infection by destabilizing the E protein	na	na	MEF	[199]
Viperin	Restrict a wide range of viruses including ZIKV	na	na	Huh7	[200]
USP38	Attenuates K48- and K63-linked polyubiquitination of E protein	na	na	HeLa/HEK293	[201]

Note: na, data not available.

7. Concluding Remarks

ZIKV infects host cells through an E protein-mediated viral entry mechanism that is a highly conserved process across humans, animals, and mosquitoes. However, the specific cell surface receptors required for viral attachment and fusion may vary among different hosts. These differences in receptor usage could contribute to variations in viral tropism and pathogenicity across species [97].

The GL, particularly the N154 glycosylation of the ZIKV envelope protein, plays a critical role in viral entry by facilitating attachment to host cell receptors and enhancing infectivity. Additionally, this glycosylation site contributes to immune evasion by modulating host immune recognition and response [105,202]. Therefore, the significant differences observed between African and Asian ZIKV lineages, particularly in glycosylation and protonation status, may influence cell tropism, viral entry, and pathogenesis in their respective hosts. The nAbs play a crucial role in neutralizing the ZIKV E protein activities during viral entry and the development of vaccines in mitigating ZIKV infection. Critical antigenic sites such as DI, DII, DIII, and combination of these domains are all major targets for nAbs. In particular, the most promising antigen site for nAb development is DIII, which not only induces potent nAbs but also causes minimal ADE [127,203,204]. Additionally, targeting the GL, the N154 glycosylation could also potentially be a promising strategy for nAb discovery because the GL is adjacent to the FL, potentially creating steric hindrance that inhibits virus entry and infection [1,143,205]. Note that although large numbers of nAbs and ZIKV E protein inhibitors and natural compounds have been reported, none of them have been approved by FDA to treat the ZIKV infection [206].

ZIKV remains a significant public health concern, particularly in regions with ongoing transmission and the potential for new outbreaks [15,207]. Given the virus's ability to spread rapidly and the devastating consequences it can develop, such as neurologically related diseases in adults and congenital malformations such as microcephaly in newborn babies or during pregnancy, continuous surveillance and research are essential. ADE may also pose a challenge for ZIKV treatment and vaccine development because cross-reactive antibodies against DENV, which co-circulates with ZIKV, can enhance ZIKV infection instead of neutralizing it, potentially leading to more severe disease [167]. The emergence of new ZIKV variants could pose additional challenges for drug discovery and vaccine development, as mutations in viral proteins may affect transmissibility, pathogenicity, and immune escape [208,209]. Variants with alterations in the E protein could potentially evade nAbs generated from previous infections reducing the effectiveness of current therapeutic and preventive measures [71]. The risk of emerging variants is heightened by factors such as international travel, urbanization, and climate change, which may expand the habitats of *Aedes* mosquitoes, the primary vectors of ZIKV [210]. In this context, the need for robust public health strategies, including monitoring of viral evolution and preparedness for vaccine updates, is more critical than ever to mitigate the impact of future ZIKV outbreaks.

Funding: This research received no external funding.

Acknowledgments: We want to thank all of those who have contributed to our understanding of the emerging ZIKV, especially those studies on molecular mechanisms underlying ZIKV–host interactions. We also want to apologize in advance if we have missed any important references that we should have included in this review. This research was funded in part by funding from the National Institute of Health (NIH R21AI129369 and R21AI175931), the U.S. Department of Veterans Affairs (VA Merit Award I01BX004652), and University of Maryland Medical Center (to R.Y.Z). The contents do not represent the views of the U.S. Department of Veterans Affairs or the United States Government.

Conflicts of Interest: The authors declare that there is no conflict of interest of any kind. The opinions expressed by the authors contributing to this journal do not necessarily reflect the opinions of the institutions with which the authors are affiliated.

Abbreviations

a.a.	Amino acids
ADE	Antibody-dependent enhancement
BBB	Blood-brain barriers
CDE	Clathrin-dependent endocytosis
CHIKV	Chikungunya virus
CIE	Clathrin-independent endocytosis
CPE	Cytopathic effect
CNS	Central nervous system
DC	Dendritic cell
DENV	Dengue virus
dsRNA	Double stranded RNA
EDE	E dimer epitope
EGCG	Epigallocatechin gallate
EM	Electron microscopy
ER	Endoplasmic reticulum
FL	Fusion loop
GBS	Guillain–Barré syndrome
GL	Glycan loop
HAVcr-1	Hepatitis A virus cellular receptor 1
hBMEC	Human brain microvascular endothelial cell
hEC	Human epithelial cell
hNPC	Human neural progenitor cell
hUVEC	Human umbilical vein endothelial cells
IFN	Interferon
IRF3	Interferon regulatory factor 3
ISGS	Interferon stimulated genes
JEV	Japanese Encephalitis virus
LAMR1	Laminin receptor 1
LR	Lateral ridge
mAb	Monoclonal antibody
MEF	Mouse Embryonic Fibroblasts
NAb	Neutralizing antibody
NPCs	Neural progenitor stem cells
NSCs	Neural Stem Cells
PAMP	Pathogen-associated molecular pattern
PGG	Pentagalloylglucose
PRRs	Pattern recognition receptors
PtdSer	Phosphatidylserine
RdRP	RNA-dependent RNA polymerase
RIG-1	RIG-1-like receptor
RLRs	RIG-I like receptors
ROS	Reactive oxygen species
sfRNA	Subgenomic flavivirus RNA
TAM	Tyro3, Axl and Mer
TGN	Trans-Golgi network
TM	Trans-membrane

TLR	Toll-like receptors
TOR	Target of rapamycin
WHO	World health organization
WNV	West Nile virus
YFV	Yellow Fever Virus
ZIKV	Zika virus

References

1. Lee, I.; Bos, S.; Li, G.; Wang, S.; Gadea, G.; Despres, P.; Zhao, R.Y. Probing Molecular Insights into Zika Virus-Host Interactions. *Viruses* **2018**, *10*, 233. [CrossRef] [PubMed]
2. Dick, G.W.; Kitchen, S.F.; Haddow, A.J. Zika Virus: A New Virus Isolated from Aedes mosquitoes in Uganda. *Nature* **1952**, *229*, 735–736.
3. Bell, T.M.; Field, E.J.; Narang, H.K. Zika virus infection of the central nervous system of mice. *Arch. Gesamte Virusforsch.* **1971**, *35*, 183–193. [CrossRef] [PubMed]
4. Dick, G.W. Zika virus. II. Pathogenicity and physical properties. *Trans. R. Soc. Trop. Med. Hyg.* **1952**, *46*, 521–534. [CrossRef]
5. Dick, G.W.; Kitchen, S.F.; Haddow, A.J. Zika virus. I. Isolations and serological specificity. *Trans. R. Soc. Trop. Med. Hyg.* **1952**, *46*, 509–520. [CrossRef]
6. Hayes, E.B. Zika virus outside Africa. *Emerg. Infect. Dis.* **2009**, *15*, 1347–1350. [CrossRef]
7. Foy, B.D.; Kobylinski, K.C.; Chilson Foy, J.L.; Blitvich, B.J.; Travassos da Rosa, A.; Haddow, A.D.; Lanciotti, R.S.; Tesh, R.B Probable non-vector-borne transmission of Zika virus, Colorado, USA. *Emerg. Infect. Dis.* **2011**, *17*, 880–882. [CrossRef]
8. Mead, P.S.; Hills, S.L.; Brooks, J.T. Zika virus as a sexually transmitted pathogen. *Curr. Opin. Infect. Dis.* **2018**, *31*, 39–44. [CrossRef]
9. Musso, D.; Nhan, T.; Robin, E.; Roche, C.; Bierlaire, D.; Zisou, K.; Shan Yan, A.; Cao-Lormeau, V.M.; Broult, J. Potential for Zika virus transmission through blood transfusion demonstrated during an outbreak in French Polynesia, November 2013 to February 2014. *Euro. Surveill.* **2014**, *19*, 20761. [CrossRef]
10. Macnamara, F.N. Zika virus: A report on three cases of human infection during an epidemic of jaundice in Nigeria. *Trans. R. Soc Trop. Med. Hyg.* **1954**, *48*, 139–145. [CrossRef]
11. Marchette, N.J.; Garcia, R.; Rudnick, A. Isolation of Zika virus from Aedes aegypti mosquitoes in Malaysia. *Am. J. Trop. Med. Hyg* **1969**, *18*, 411–415. [CrossRef] [PubMed]
12. Olson, J.G.; Ksiazek, T.G.; Suhandiman; Triwibowo. Zika virus, a cause of fever in Central Java, Indonesia. *Trans. R. Soc. Trop Med. Hyg.* **1981**, *75*, 389–393. [CrossRef] [PubMed]
13. Darwish, M.A.; Hoogstraal, H.; Roberts, T.J.; Ahmed, I.P.; Omar, F. A sero-epidemiological survey for certain arboviruses (Togaviridae) in Pakistan. *Trans. R. Soc. Trop. Med. Hyg.* **1983**, *77*, 442–445. [CrossRef] [PubMed]
14. Duffy, M.R.; Chen, T.H.; Hancock, W.T.; Powers, A.M.; Kool, J.L.; Lanciotti, R.S.; Pretrick, M.; Marfel, M.; Holzbauer, S.; Dubray C.; et al. Zika virus outbreak on Yap Island, Federated States of Micronesia. *N. Engl. J. Med.* **2009**, *360*, 2536–2543. [CrossRef]
15. Gulland, A. Zika virus is a global public health emergency, declares WHO. *BMJ* **2016**, *352*, i657. [CrossRef]
16. Heukelbach, J.; Alencar, C.H.; Kelvin, A.A.; de Oliveira, W.K.; Pamplona de Goes Cavalcanti, L. Zika virus outbreak in Brazil. *J. Infect. Dev. Ctries.* **2016**, *10*, 116–120. [CrossRef]
17. Yadav, P.D.; Niyas, V.K.M.; Arjun, R.; Sahay, R.R.; Shete, A.M.; Sapkal, G.N.; Pawar, S.D.; Patil, D.Y.; Gupta, N.; Abraham, P. Detection of Zika virus disease in Thiruvananthapuram, Kerala, India 2021 during the second wave of COVID-19 pandemic. *J. Med. Virol.* **2022**, *94*, 2346. [CrossRef]
18. Yu, X.; Cheng, G. Adaptive Evolution as a Driving Force of the Emergence and Re-Emergence of Mosquito-Borne Viral Disease *Viruses* **2022**, *14*, 435. [CrossRef]
19. Chambers, T.J.; Hahn, C.S.; Galler, R.; Rice, C.M. Flavivirus genome organization, expression, and replication. *Annu. Rev Microbiol.* **1990**, *44*, 649–688. [CrossRef]
20. Sirohi, D.; Chen, Z.; Sun, L.; Klose, T.; Pierson, T.C.; Rossmann, M.G.; Kuhn, R.J. The 3.8 A resolution cryo-EM structure of Zika virus. *Science* **2016**, *352*, 467–470. [CrossRef]
21. Harris, E.; Holden, K.L.; Edgil, D.; Polacek, C.; Clyde, K. Molecular biology of flaviviruses. *Novartis Found. Symp.* **2006**, *277*, 23–3 discussion 40, 71–23, 251–253. [PubMed]
22. Li, G.; Poulsen, M.; Fenyvuesvolgyi, C.; Yashiroda, Y.; Yoshida, M.; Simard, J.M.; Gallo, R.C.; Zhao, R.Y. Characterization of cytopathic factors through genome-wide analysis of the Zika viral proteins in fission yeast. *Proc. Natl. Acad. Sci. USA* **201** *114*, E376–E385. [CrossRef] [PubMed]
23. Yuan, L.; Huang, X.-Y.; Liu, Z.-Y.; Zhang, F.; Zhu, X.-L.; Yu, J.-Y.; Ji, X.; Xu, Y.-P.; Li, G.; Li, C.J.S. A single mutation in the prM protein of Zika virus contributes to fetal microcephaly. *Science* **2017**, *358*, 933–936. [CrossRef] [PubMed]
24. Nambala, P.; Su, W.-C.J.F.i.m. Role of Zika virus prM protein in viral pathogenicity and use in vaccine development. *Fron Microbiol.* **2018**, *9*, 1797. [CrossRef]

25. Amberg, S.M.; Nestorowicz, A.; McCourt, D.W.; Rice, C.M. NS2B-3 proteinase-mediated processing in the yellow fever virus structural region: In vitro and in vivo studies. *J. Virol.* **1994**, *68*, 3794–3802. [CrossRef]
26. Lobigs, M.; Lee, E.; Ng, M.L.; Pavy, M.; Lobigs, P. A flavivirus signal peptide balances the catalytic activity of two proteases and thereby facilitates virus morphogenesis. *Virology* **2010**, *401*, 80–89. [CrossRef]
27. Li, G.; Bos, S.; Tsetsarkin, K.A.; Pletnev, A.G.; Despres, P.; Gadea, G.; Zhao, R.Y. The Roles of prM-E Proteins in Historical and Epidemic Zika Virus-mediated Infection and Neurocytotoxicity. *Viruses* **2019**, *11*, 157. [CrossRef]
28. Stadler, K.; Allison, S.L.; Schalich, J.; Heinz, F.X. Proteolytic activation of tick-borne encephalitis virus by furin. *J. Virol.* **1997**, *71*, 8475–8481. [CrossRef]
29. Elshuber, S.; Allison, S.L.; Heinz, F.X.; Mandl, C.W. Cleavage of protein prM is necessary for infection of BHK-21 cells by tick-borne encephalitis virus. *J. Gen. Virol.* **2003**, *84*, 183–191. [CrossRef]
30. Majowicz, S.A.; Narayanan, A.; Moustafa, I.M.; Bator, C.M.; Hafenstein, S.L.; Jose, J.J.n.V. Zika virus M protein latches and locks the E protein from transitioning to an immature state after prM cleavage. *NPJ Viruses* **2023**, *1*, 4. [CrossRef]
31. Faye, O.; Freire, C.C.; Iamarino, A.; Faye, O.; de Oliveira, J.V.; Diallo, M.; Zanotto, P.M.; Sall, A.A. Molecular evolution of Zika virus during its emergence in the 20(th) century. *PLoS Negl. Trop. Dis.* **2014**, *8*, e2636. [CrossRef] [PubMed]
32. Sundar, S.; Piramanayagam, S.; Natarajan, J.J.V.G. A review on structural genomics approach applied for drug discovery against three vector-borne viral diseases: Dengue, Chikungunya and Zika. *Virus Genes* **2022**, *58*, 151–171. [CrossRef] [PubMed]
33. Lei, J.; Hansen, G.; Nitsche, C.; Klein, C.D.; Zhang, L.; Hilgenfeld, R.J.S. Crystal structure of Zika virus NS2B-NS3 protease in complex with a boronate inhibitor. *Science* **2016**, *353*, 503–505. [CrossRef] [PubMed]
34. Guo, M.; Hui, L.; Nie, Y.; Tefsen, B.; Wu, Y.J.S.C.L.S. ZIKV viral proteins and their roles in virus-host interactions. *Sci. China Life Sci.* **2021**, *64*, 709–719. [CrossRef]
35. Cortese, M.; Goellner, S.; Acosta, E.G.; Neufeldt, C.J.; Oleksiuk, O.; Lampe, M.; Haselmann, U.; Funaya, C.; Schieber, N.; Ronchi, P.J.C.r. Ultrastructural characterization of Zika virus replication factories. *Cell Rep.* **2017**, *18*, 2113–2123. [CrossRef]
36. Tabata, T.; Petitt, M.; Puerta-Guardo, H.; Michlmayr, D.; Wang, C.; Fang-Hoover, J.; Harris, E.; Pereira, L. Zika Virus Targets Different Primary Human Placental Cells, Suggesting Two Routes for Vertical Transmission. *Cell Host Microbe* **2016**, *20*, 155–166. [CrossRef]
37. de Paula Freitas, B.; de Oliveira Dias, J.R.; Prazeres, J.; Sacramento, G.A.; Ko, A.I.; Maia, M.; Belfort, R., Jr. Ocular Findings in Infants With Microcephaly Associated With Presumed Zika Virus Congenital Infection in Salvador, Brazil. *JAMA Ophthalmol.* **2016**, *134*, 529–535. [CrossRef]
38. Sager, G.; Gabaglio, S.; Sztul, E.; Belov, G.A.J.V. Role of host cell secretory machinery in Zika virus life cycle. *Viruses* **2018**, *10*, 559. [CrossRef]
39. Mukhopadhyay, S.; Kuhn, R.J.; Rossmann, M.G. A structural perspective of the flavivirus life cycle. *Nat. Rev. Microbiol.* **2005**, *3*, 13–22. [CrossRef]
40. Zhang, X.; Jia, R.; Shen, H.; Wang, M.; Yin, Z.; Cheng, A. Structures and functions of the envelope glycoprotein in flavivirus infections. *Viruses* **2017**, *9*, 338. [CrossRef]
41. Shi, Y.; Gao, G.F. Structural biology of the Zika virus. *Trends Biochem. Sci.* **2017**, *42*, 443–456. [CrossRef] [PubMed]
42. Barrows, N.J.; Campos, R.K.; Liao, K.-C.; Prasanth, K.R.; Soto-Acosta, R.; Yeh, S.-C.; Schott-Lerner, G.; Pompon, J.; Sessions, O.M.; Bradrick, S.S. Biochemistry and molecular biology of flaviviruses. *Chem. Rev.* **2018**, *118*, 4448–4482. [CrossRef] [PubMed]
43. Frumence, E.; Haddad, J.G.; Vanwalscappel, B.; Andries, J.; Decotter, J.; Viranaicken, W.; Gadea, G.; Desprès, P. Immune Reactivity of a 20-mer Peptide Representing the Zika E Glycan Loop Involves the Antigenic Determinants E-152/156/158. *Viruses* **2020**, *12*, 1258. [CrossRef] [PubMed]
44. Dai, L.; Song, J.; Lu, X.; Deng, Y.Q.; Musyoki, A.M.; Cheng, H.; Zhang, Y.; Yuan, Y.; Song, H.; Haywood, J.; et al. Structures of the Zika Virus Envelope Protein and Its Complex with a Flavivirus Broadly Protective Antibody. *Cell Host Microbe* **2016**, *19*, 696–704. [CrossRef]
45. Routhu, N.K.; Lehoux, S.D.; Rouse, E.A.; Bidokhti, M.R.M.; Giron, L.B.; Anzurez, A.; Reid, S.P.; Abdel-Mohsen, M.; Cummings, R.D.; Byrareddy, S.N. Glycosylation of Zika Virus is Important in Host-Virus Interaction and Pathogenic Potential. *Int. J. Mol. Sci.* **2019**, *20*, 5206. [CrossRef]
46. Hu, T.; Wu, Z.; Wu, S.; Chen, S.; Cheng, A. The key amino acids of E protein involved in early flavivirus infection: Viral entry. *Virol. J.* **2021**, *18*, 136. [CrossRef]
47. Chakraborty, S. Computational analysis of perturbations in the post-fusion Dengue virus envelope protein highlights known epitopes and conserved residues in the Zika virus. *F1000Research* **2016**, *5*, 1150. [CrossRef]
48. Bressanelli, S.; Stiasny, K.; Allison, S.L.; Stura, E.A.; Duquerroy, S.; Lescar, J.; Heinz, F.X.; Rey, F.A. Structure of a flavivirus envelope glycoprotein in its low-pH-induced membrane fusion conformation. *EMBO J.* **2004**, *23*, 728–738. [CrossRef]
49. Lee, J.W.-M.; Chu, J.J.-H.; Ng, M.-L. Quantifying the specific binding between West Nile virus envelope domain III protein and the cellular receptor $\alpha V\beta 3$ integrin. *J. Biol. Chem.* **2006**, *281*, 1352–1360. [CrossRef]

50. Agrelli, A.; de Moura, R.R.; Crovella, S.; Brandão, L.A.C. ZIKA virus entry mechanisms in human cells. *Infect. Genet. Evol.* **2019**, *69*, 22–29. [CrossRef]
51. Qu, P.; Zhang, W.; Li, D.; Zhang, C.; Liu, Q.; Zhang, X.; Wang, X.; Dai, W.; Xu, Y.; Leng, Q. Insect cell-produced recombinan protein subunit vaccines protect against Zika virus infection. *Antivir. Res.* **2018**, *154*, 97–103. [CrossRef] [PubMed]
52. Shin, M.; Kim, K.; Lee, H.J.; Jung, Y.J.; Park, J.; Hahn, T.W. Vaccination with a Zika virus envelope domain III protein induce neutralizing antibodies and partial protection against Asian genotype in immunocompetent mice. *Trop. Med. Health* **2022**, *50*, 91 [CrossRef]
53. Gallichotte, E.N.; Dinnon, K.H., 3rd; Lim, X.N.; Ng, T.S.; Lim, E.X.Y.; Menachery, V.D.; Lok, S.M.; Baric, R.S. CD-loop Extension in Zika Virus Envelope Protein Key for Stability and Pathogenesis. *J. Infect. Dis.* **2017**, *216*, 1196–1204. [CrossRef] [PubMed]
54. Carbaugh, D.L.; Lazear, H.M. Flavivirus envelope protein glycosylation: Impacts on viral infection and pathogenesis. *J. Viro* **2020**, *94*, 10–1128. [CrossRef] [PubMed]
55. Feng, T.; Zhang, J.; Chen, Z.; Pan, W.; Chen, Z.; Yan, Y.; Dai, J. Glycosylation of viral proteins: Implication in virus–host interactio and virulence. *Virulence* **2022**, *13*, 670–683. [CrossRef]
56. Carbaugh, D.L.; Baric, R.S.; Lazear, H.M. Envelope Protein Glycosylation Mediates Zika Virus Pathogenesis. *J. Virol.* **201** *93*, e00113-19. [CrossRef]
57. Bos, S.; Viranaicken, W.; Frumence, E.; Li, G.; Despres, P.; Zhao, R.Y.; Gadea, G. The Envelope Residues E152/156/158 of Zik Virus Influence the Early Stages of Virus Infection in Human Cells. *Cells* **2019**, *8*, 1444. [CrossRef]
58. Lei, Y.; Yu, H.; Dong, Y.; Yang, J.; Ye, W.; Wang, Y.; Chen, W.; Jia, Z.; Xu, Z.; Li, Z.J.P.O. Characterization of N-glycan structures o the surface of mature dengue 2 virus derived from insect cells. *PLoS ONE* **2015**, *10*, e0132122. [CrossRef]
59. Pralow, A.; Nikolay, A.; Leon, A.; Genzel, Y.; Rapp, E.; Reichl, U.J.S.R. Site-specific N-glycosylation analysis of animal ce culture-derived Zika virus proteins. *Sci. Rep.* **2021**, *11*, 5147. [CrossRef]
60. Gong, D.; Zhang, T.H.; Zhao, D.; Du, Y.; Chapa, T.J.; Shi, Y.; Wang, L.; Contreras, D.; Zeng, G.; Shi, P.Y.; et al. High-Throughpu Fitness Profiling of Zika Virus E Protein Reveals Different Roles for Glycosylation during Infection of Mammalian and Mosquit Cells. *iScience* **2018**, *1*, 97–111. [CrossRef]
61. Sun, J.; Li, Y.; Liu, P.; Lin, J. Study of the mechanism of protonated histidine-induced conformational changes in the Zika viru dimeric envelope protein using accelerated molecular dynamic simulations. *J. Mol. Graph. Model.* **2017**, *74*, 203–214. [CrossRe [PubMed]
62. Moller-Tank, S.; Maury, W. Phosphatidylserine receptors: Enhancers of enveloped virus entry and infection. *Virology* **201** *468–470*, 565–580. [CrossRef] [PubMed]
63. Song, D.H.; Garcia, G., Jr.; Situ, K.; Chua, B.A.; Hong, M.L.O.; Do, E.A.; Ramirez, C.M.; Harui, A.; Arumugaswami, V.; Morizono, Development of a blocker of the universal phosphatidylserine- and phosphatidylethanolamine-dependent viral entry pathwa *Virology* **2021**, *560*, 17–33. [CrossRef] [PubMed]
64. Reddy, T.; Sansom, M.S. The Role of the Membrane in the Structure and Biophysical Robustness of the Dengue Virion Envelop *Structure* **2016**, *24*, 375–382. [CrossRef]
65. Ghosh Roy, S. TAM receptors: A phosphatidylserine receptor family and its implications in viral infections. *Int. Rev. Cell M. Biol.* **2020**, *357*, 81–122. [CrossRef]
66. Tang, D.; Wang, Y.; Dong, X.; Yuan, Y.; Kang, F.; Tian, W.; Wang, K.; Li, H.; Qi, S. Scramblases and virus infection. *BioEssays* **202** *44*, 2100261. [CrossRef]
67. Chua, B.A.; Ngo, J.A.; Situ, K.; Morizono, K. Roles of phosphatidylserine exposed on the viral envelope and cell membrane HIV-1 replication. *Cell Commun. Signal.* **2019**, *17*, 132. [CrossRef]
68. Berneck, B.S.; Rockstroh, A.; Fertey, J.; Grunwald, T.; Ulbert, S.J.V. A recombinant Zika virus envelope protein with mutations the conserved fusion loop leads to reduced antibody cross-reactivity upon vaccination. *Vaccines* **2020**, *8*, 603. [CrossRef]
69. Kikawa, C.; Cartwright-Acar, C.H.; Stuart, J.B.; Contreras, M.; Levoir, L.M.; Evans, M.J.; Bloom, J.D.; Goo, L. The effect of sing mutations in Zika virus envelope on escape from broadly neutralizing antibodies. *J. Virol.* **2023**, *97*, e0141423. [CrossRef]
70. Ma, X.; Yuan, Z.; Yi, Z. Identification and characterization of key residues in Zika virus envelope protein for virus assembly a entry. *Emerg. Microbes Infect.* **2022**, *11*, 1604–1620. [CrossRef]
71. Sourisseau, M.; Lawrence, D.J.P.; Schwarz, M.C.; Storrs, C.H.; Veit, E.C.; Bloom, J.D.; Evans, M.J. Deep mutational scanni comprehensively maps how Zika envelope protein mutations affect viral growth and antibody escape. *J. Virol.* **2019**, *93*, 10–11 [CrossRef] [PubMed]
72. Esser-Nobis, K.; Aarreberg, L.D.; Roby, J.A.; Fairgrieve, M.R.; Green, R.; Gale, M., Jr. Comparative Analysis of African and Asi Lineage-Derived Zika Virus Strains Reveals Differences in Activation of and Sensitivity to Antiviral Innate Immunity. *J. Vi* **2019**, *93*, e00640-19. [CrossRef] [PubMed]
73. Shao, Q.; Herrlinger, S.; Zhu, Y.N.; Yang, M.; Goodfellow, F.; Stice, S.L.; Qi, X.P.; Brindley, M.A.; Chen, J.F. The African Zika vi MR-766 is more virulent and causes more severe brain damage than current Asian lineage and dengue virus. *Development* **20** *144*, 4114–4124. [CrossRef] [PubMed]

74. Anfasa, F.; Siegers, J.Y.; van der Kroeg, M.; Mumtaz, N.; Stalin Raj, V.; de Vrij, F.M.S.; Widagdo, W.; Gabriel, G.; Salinas, S.; Simonin, Y.; et al. Phenotypic Differences between Asian and African Lineage Zika Viruses in Human Neural Progenitor Cells. *mSphere* **2017**, *2*, e00292-17. [CrossRef]
75. Liu, Y.; Liu, J.; Du, S.; Shan, C.; Nie, K.; Zhang, R.; Li, X.F.; Zhang, R.; Wang, T.; Qin, C.F.; et al. Evolutionary enhancement of Zika virus infectivity in Aedes aegypti mosquitoes. *Nature* **2017**, *545*, 482–486. [CrossRef]
76. Sievers, F.; Wilm, A.; Dineen, D.; Gibson, T.J.; Karplus, K.; Li, W.; Lopez, R.; McWilliam, H.; Remmert, M.; Soding, J.; et al. Fast, scalable generation of high-quality protein multiple sequence alignments using Clustal Omega. *Mol. Syst. Biol.* **2011**, *7*, 539. [CrossRef]
77. Bianchini, G.; Sanchez-Baracaldo, P. TreeViewer: Flexible, modular software to visualise and manipulate phylogenetic trees. *Ecol. Evol.* **2024**, *14*, e10873. [CrossRef]
78. Faria, N.R.; Azevedo Rdo, S.; Kraemer, M.U.; Souza, R.; Cunha, M.S.; Hill, S.C.; Theze, J.; Bonsall, M.B.; Bowden, T.A.; Rissanen, I.; et al. Zika virus in the Americas: Early epidemiological and genetic findings. *Science* **2016**, *352*, 345–349. [CrossRef]
79. Haddow, A.D.; Schuh, A.J.; Yasuda, C.Y.; Kasper, M.R.; Heang, V.; Huy, R.; Guzman, H.; Tesh, R.B.; Weaver, S.C. Genetic characterization of Zika virus strains: Geographic expansion of the Asian lineage. *PLoS Negl. Trop. Dis.* **2012**, *6*, e1477. [CrossRef]
80. Beaver, J.T.; Lelutiu, N.; Habib, R.; Skountzou, I. Evolution of Two Major Zika Virus Lineages: Implications for Pathology, Immune Response, and Vaccine Development. *Front. Immunol.* **2018**, *9*, 1640. [CrossRef]
81. Bos, S.; Viranaicken, W.; Turpin, J.; El-Kalamouni, C.; Roche, M.; Krejbich-Trotot, P.; Despres, P.; Gadea, G. The structural proteins of epidemic and historical strains of Zika virus differ in their ability to initiate viral infection in human host cells. *Virology* **2018**, *516*, 265–273. [CrossRef] [PubMed]
82. Frumence, E.; Viranaicken, W.; Bos, S.; Alvarez-Martinez, M.T.; Roche, M.; Arnaud, J.D.; Gadea, G.; Despres, P. A Chimeric Zika Virus between Viral Strains MR766 and BeH819015 Highlights a Role for E-glycan Loop in Antibody-mediated Virus Neutralization. *Vaccines* **2019**, *7*, 55. [CrossRef] [PubMed]
83. Ramaiah, A.; Dai, L.; Contreras, D.; Sinha, S.; Sun, R.; Arumugaswami, V. Comparative analysis of protein evolution in the genome of pre-epidemic and epidemic Zika virus. *Infect. Genet. Evol.* **2017**, *51*, 74–85. [CrossRef]
84. Goo, L.; DeMaso, C.R.; Pelc, R.S.; Ledgerwood, J.E.; Graham, B.S.; Kuhn, R.J.; Pierson, T.C. The Zika virus envelope protein glycan loop regulates virion antigenicity. *Virology* **2018**, *515*, 191–202. [CrossRef] [PubMed]
85. Annamalai, A.S.; Pattnaik, A.; Sahoo, B.R.; Muthukrishnan, E.; Natarajan, S.K.; Steffen, D.; Vu, H.L.X.; Delhon, G.; Osorio, F.A.; Petro, T.M.; et al. Zika Virus Encoding Non-Glycosylated Envelope Protein is Attenuated and Defective in Neuroinvasion. *J. Virol.* **2017**, *91*, e01348-17. [CrossRef]
86. Bowen, J.R.; Zimmerman, M.G.; Suthar, M.S. Taking the defensive: Immune control of Zika virus infection. *Virus Res.* **2018**, *254*, 21–26. [CrossRef]
87. Mesci, P.; Macia, A.; LaRock, C.N.; Tejwani, L.; Fernandes, I.R.; Suarez, N.A.; Zanotto, P.M.d.A.; Beltrão-Braga, P.C.B.; Nizet, V.; Muotri, A.R. Modeling neuro-immune interactions during Zika virus infection. *Hum. Mol. Genet.* **2018**, *27*, 41–52. [CrossRef]
88. Sheridan, M.A.; Yunusov, D.; Balaraman, V.; Alexenko, A.P.; Yabe, S.; Verjovski-Almeida, S.; Schust, D.J.; Franz, A.W.; Sadovsky, Y.; Ezashi, T. Vulnerability of primitive human placental trophoblast to Zika virus. *Proc. Natl. Acad. Sci. USA* **2017**, *114*, E1587–E1596. [CrossRef]
89. Li, C.; Xu, D.; Ye, Q.; Hong, S.; Jiang, Y.; Liu, X.; Zhang, N.; Shi, L.; Qin, C.F.; Xu, Z. Zika Virus Disrupts Neural Progenitor Development and Leads to Microcephaly in Mice. *Cell Stem Cell* **2016**, *19*, 672. [CrossRef]
90. Nowakowski, T.J.; Pollen, A.A.; Di Lullo, E.; Sandoval-Espinosa, C.; Bershteyn, M.; Kriegstein, A.R. Expression Analysis Highlights AXL as a Candidate Zika Virus Entry Receptor in Neural Stem Cells. *Cell Stem Cell* **2016**, *18*, 591–596. [CrossRef]
91. Meertens, L.; Labeau, A.; Dejarnac, O.; Cipriani, S.; Sinigaglia, L.; Bonnet-Madin, L.; Le Charpentier, T.; Hafirassou, M.L.; Zamborlini, A.; Cao-Lormeau, V.-M. Axl mediates ZIKA virus entry in human glial cells and modulates innate immune responses. *Cell Rep.* **2017**, *18*, 324–333. [CrossRef] [PubMed]
92. Zwernik, S.D.; Adams, B.H.; Raymond, D.A.; Warner, C.M.; Kassam, A.B.; Rovin, R.A.; Akhtar, P. AXL receptor is required for Zika virus strain MR-766 infection in human glioblastoma cell lines. *Mol. Ther. Oncolytics* **2021**, *23*, 447–457. [CrossRef] [PubMed]
93. Al Kafri, N.; Ahnström, J.; Teraz-Orosz, A.; Chaput, L.; Singh, N.; Villoutreix, B.O.; Hafizi, S. The first laminin G-like domain of protein S is essential for binding and activation of Tyro3 receptor and intracellular signalling. *Biochem. Biophys. Rep.* **2022**, *30*, 101263. [CrossRef] [PubMed]
94. Lemke, G.; Burstyn-Cohen, T. TAM receptors and the clearance of apoptotic cells. *Ann. N. Y. Acad. Sci.* **2010**, *1209*, 23–29. [CrossRef]
95. Espino, A.; Gouilly, J.; Chen, Q.; Colin, P.; Guerby, P.; Izopet, J.; Amara, A.; Tabiasco, J.; Al-Daccak, R.; El Costa, H. The mechanisms underlying the immune control of zika virus infection at the maternal-fetal interface. *Front. Immunol.* **2022**, *13*, 1000861. [CrossRef]
96. Komarasamy, T.V.; Adnan, N.A.A.; James, W.; Balasubramaniam, V. Zika Virus Neuropathogenesis: The Different Brain Cells, Host Factors and Mechanisms Involved. *Front. Immunol.* **2022**, *13*, 773191. [CrossRef]

97. Hamel, R.; Dejarnac, O.; Wichit, S.; Ekchariyawat, P.; Neyret, A.; Luplertlop, N.; Perera-Lecoin, M.; Surasombatpattana, P.; Talignani, L.; Thomas, F. Biology of Zika virus infection in human skin cells. *J. Virol.* **2015**, *89*, 8880–8896. [CrossRef]
98. Richard, A.S.; Shim, B.-S.; Kwon, Y.-C.; Zhang, R.; Otsuka, Y.; Schmitt, K.; Berri, F.; Diamond, M.S.; Choe, H. AXL-dependent infection of human fetal endothelial cells distinguishes Zika virus from other pathogenic flaviviruses. *Proc. Natl. Acad. Sci. USA* **2017**, *114*, 2024–2029. [CrossRef]
99. Moller-Tank, S.; Kondratowicz, A.S.; Davey, R.A.; Rennert, P.D.; Maury, W. Role of the phosphatidylserine receptor TIM-1 in enveloped-virus entry. *J. Virol.* **2013**, *87*, 8327–8341. [CrossRef]
100. Estevez-Herrera, J.; Perez-Yanes, S.; Cabrera-Rodriguez, R.; Marquez-Arce, D.; Trujillo-Gonzalez, R.; Machado, J.D.; Madrid, R.; Valenzuela-Fernandez, A. Zika Virus Pathogenesis: A Battle for Immune Evasion. *Vaccines* **2021**, *9*, 294. [CrossRef]
101. Eder, J.; Zijlstra-Willems, E.; Koen, G.; Kootstra, N.A.; Wolthers, K.C.; Geijtenbeek, T.B.J.F.i.I. Transmission of Zika virus by dendritic cell subsets in skin and vaginal mucosa. *Cell* **2023**, *14*, 1125565. [CrossRef] [PubMed]
102. Jaimipuk, T.; Sachdev, S.; Yoksan, S.; Thepparit, C. A Small-Plaque Isolate of the Zika Virus with Envelope Domain III Mutations Affect Viral Entry and Replication in Mammalian but Not Mosquito Cells. *Viruses* **2022**, *14*, 480. [CrossRef] [PubMed]
103. Chakraborty, S. MEPPitope: Spatial, electrostatic and secondary structure perturbations in the post-fusion Dengue virus envelope protein highlights known epitopes and conserved residues in the Zika virus. *F1000Research* **2016**, *5*, 1150. [CrossRef] [PubMed]
104. Sirohi, D.; Kuhn, R.J. Zika virus structure, maturation, and receptors. *J. Infect. Dis.* **2017**, *216*, S935–S944. [CrossRef]
105. Fontes-Garfias, C.R.; Shan, C.; Luo, H.; Muruato, A.E.; Medeiros, D.B.A.; Mays, E.; Xie, X.; Zou, J.; Roundy, C.M.; Wakamiya, M Functional analysis of glycosylation of Zika virus envelope protein. *Cell Rep.* **2017**, *21*, 1180–1190. [CrossRef]
106. Mossenta, M.; Marchese, S.; Poggianella, M.; Slon Campos, J.L.; Burrone, O.R. Role of N-glycosylation on Zika virus E protein secretion, viral assembly and infectivity. *Biochem. Biophys. Res. Commun.* **2017**, *492*, 579–586. [CrossRef]
107. Morizono, K.; Chen, I.S. Role of phosphatidylserine receptors in enveloped virus infection. *J. Virol.* **2014**, *88*, 4275–4290. [CrossRef]
108. Perera-Lecoin, M.; Meertens, L.; Carnec, X.; Amara, A. Flavivirus entry receptors: An update. *Viruses* **2013**, *6*, 69–88. [CrossRef]
109. Li, M.; Zhang, D.; Li, C.; Zheng, Z.; Fu, M.; Ni, F.; Liu, Y.; Du, T.; Wang, H.; Griffin, G.E.J.F.I.M. Characterization of Zika virus endocytic pathways in human glioblastoma cells. *Front. Microbiol.* **2020**, *11*, 242. [CrossRef]
110. Owczarek, K.; Chykunova, Y.; Jassoy, C.; Maksym, B.; Rajfur, Z.; Pyrc, K. Zika virus: Mapping and reprogramming the entry. *Cel Commun. Signal.* **2019**, *17*, 41. [CrossRef]
111. Rinkenberger, N.; Schoggins, J.W. Comparative analysis of viral entry for Asian and African lineages of Zika virus. *Virology* **2019** *533*, 59–67. [CrossRef] [PubMed]
112. Li, Y.; Wang, X.; Bi, S.; Zhao, K.; Yu, C. Inhibition of Mer and Axl receptor tyrosine kinases leads to increased apoptosis an improved chemosensitivity in human neuroblastoma. *Biochem. Biophys. Res. Commun.* **2015**, *457*, 461–466. [CrossRef] [PubMed
113. Tappe, D.; Perez-Giron, J.V.; Zammarchi, L.; Rissland, J.; Ferreira, D.F.; Jaenisch, T.; Gomez-Medina, S.; Gunther, S.; Bartoloni, A Munoz-Fontela, C.; et al. Cytokine kinetics of Zika virus-infected patients from acute to reconvalescent phase. *Med. Microbio Immunol.* **2016**, *205*, 269–273. [CrossRef] [PubMed]
114. Galliez, R.M.; Spitz, M.; Rafful, P.P.; Cagy, M.; Escosteguy, C.; Germano, C.S.; Sasse, E.; Goncalves, A.L.; Silveira, P.P.; Pezzutc P.; et al. Zika Virus Causing Encephalomyelitis Associated With Immunoactivation. *Open Forum Infect. Dis.* **2016**, *3*, ofw20 [CrossRef]
115. Ornelas Pereira, I.; Santelli, A.; Leite, P.L.; Attell, J.; Bertolli, J.; Kotzky, K.; Araujo, W.N.; Peacock, G. Parental Stress in Primar Caregivers of Children with Evidence of Congenital Zika Virus Infection in Northeastern Brazil. *Matern. Child Health J.* **202** *25*, 360–367. [CrossRef]
116. Figueiredo, C.P.; Barros-Aragao, F.G.Q.; Neris, R.L.S.; Frost, P.S.; Soares, C.; Souza, I.N.O.; Zeidler, J.D.; Zamberlan, D.C.; d Sousa, V.L.; Souza, A.S.; et al. Zika virus replicates in adult human brain tissue and impairs synapses and memory in mice. *Na Commun.* **2019**, *10*, 3890. [CrossRef]
117. Yang, M.; Sun, H.; Lai, H.; Hurtado, J.; Chen, Q. Plant-produced Zika virus envelope protein elicits neutralizing immun responses that correlate with protective immunity against Zika virus in mice. *Plant. Biotechnol. J.* **2018**, *16*, 572–580. [CrossRef]
118. Shin, M.; Kim, K.; Lee, H.J.; Lee, R.; Jung, Y.J.; Park, J.; Hahn, T.W. Zika virus baculovirus-expressed envelope protein elicite humoral and cellular immunity in immunocompetent mice. *Sci. Rep.* **2022**, *12*, 660. [CrossRef]
119. Lunardelli, V.A.S.; de Souza Apostolico, J.; Souza, H.F.S.; Coirada, F.C.; Martinho, J.A.; Astray, R.M.; Boscardin, S.B.; Ros D.S. ZIKV-envelope proteins induce specific humoral and cellular immunity in distinct mice strains. *Sci. Rep.* **2022**, *12*, 1573 [CrossRef]
120. Hu, H.; Feng, Y.; He, M.L. Targeting Type I Interferon Induction and Signaling: How Zika Virus Escapes from Host Innat Immunity. *Int. J. Biol. Sci.* **2023**, *19*, 3015–3028. [CrossRef]
121. Anglero-Rodriguez, Y.I.; MacLeod, H.J.; Kang, S.; Carlson, J.S.; Jupatanakul, N.; Dimopoulos, G. Aedes aegypti Molecul Responses to Zika Virus: Modulation of Infection by the Toll and Jak/Stat Immune Pathways and Virus Host Factors. *Fror Microbiol.* **2017**, *8*, 2050. [CrossRef]

22. Strange, D.P.; Jiyarom, B.; Pourhabibi Zarandi, N.; Xie, X.; Baker, C.; Sadri-Ardekani, H.; Shi, P.-Y.; Verma, S. Axl promotes zika virus entry and modulates the antiviral state of human sertoli cells. *mBio* **2019**, *10*, e01372-19. [CrossRef] [PubMed]
23. Huang, G.G.; Wang, H.Y.; Wang, X.H.; Yang, T.; Zhang, X.M.; Feng, C.L.; Zhao, W.M.; Tang, W. Atranorin inhibits Zika virus infection in human glioblastoma cell line SNB-19 via targeting Zika virus envelope protein. *Phytomedicine* **2024**, *125*, 155343. [CrossRef] [PubMed]
24. Mellors, J.; Tipton, T.; Longet, S.; Carroll, M. Viral Evasion of the Complement System and Its Importance for Vaccines and Therapeutics. *Front. Immunol.* **2020**, *11*, 1450. [CrossRef] [PubMed]
25. Malekshahi, Z.; Schiela, B.; Bernklau, S.; Banki, Z.; Wurzner, R.; Stoiber, H. Interference of the Zika Virus E-Protein With the Membrane Attack Complex of the Complement System. *Front. Immunol.* **2020**, *11*, 569549. [CrossRef]
26. Wen, D.; Li, S.; Dong, F.; Zhang, Y.; Lin, Y.; Wang, J.; Zou, Z.; Zheng, A. N-glycosylation of viral E protein is the determinant for vector midgut invasion by flaviviruses. *mBio* **2018**, *9*, e00046-18. [CrossRef]
27. Dussupt, V.; Modjarrad, K.; Krebs, S.J. Landscape of monoclonal antibodies targeting zika and dengue: Therapeutic solutions and critical insights for vaccine development. *Front. Immunol.* **2021**, *11*, 621043. [CrossRef]
28. Rey, F.A.; Stiasny, K.; Vaney, M.C.; Dellarole, M.; Heinz, F.X. The bright and the dark side of human antibody responses to flaviviruses: Lessons for vaccine design. *EMBO Rep.* **2018**, *19*, 206–224. [CrossRef]
29. Gallichotte, E.N.; Young, E.F.; Baric, T.J.; Yount, B.L.; Metz, S.W.; Begley, M.C.; de Silva, A.M.; Baric, R.S. Role of Zika Virus Envelope Protein Domain III as a Target of Human Neutralizing Antibodies. *mBio* **2019**, *10*, e01485-19. [CrossRef]
30. Yang, C.; Gong, R.; de Val, N. Development of Neutralizing Antibodies against Zika Virus Based on Its Envelope Protein Structure. *Virol. Sin.* **2019**, *34*, 168–174. [CrossRef]
31. Zhao, H.; Xu, L.; Bombardi, R.; Nargi, R.; Deng, Z.; Errico, J.M.; Nelson, C.A.; Dowd, K.A.; Pierson, T.C.; Crowe, J.E.; et al. Mechanism of differential Zika and dengue virus neutralization by a public antibody lineage targeting the DIII lateral ridge. *J. Exp. Med.* **2020**, *217*, e20191792. [CrossRef] [PubMed]
32. Zhao, H.; Fernandez, E.; Dowd, K.A.; Speer, S.D.; Platt, D.J.; Gorman, M.J.; Govero, J.; Nelson, C.A.; Pierson, T.C.; Diamond, M.S. Structural basis of Zika virus-specific antibody protection. *Cell* **2016**, *166*, 1016–1027. [CrossRef] [PubMed]
33. Van Rompay, K.K.A.; Coffey, L.L.; Kapoor, T.; Gazumyan, A.; Keesler, R.I.; Jurado, A.; Peace, A.; Agudelo, M.; Watanabe, J.; Usachenko, J. A combination of two human monoclonal antibodies limits fetal damage by Zika virus in macaques. *Proc. Natl. Acad. Sci. USA* **2020**, *117*, 7981–7989. [CrossRef] [PubMed]
34. Robbiani, D.F.; Bozzacco, L.; Keeffe, J.R.; Khouri, R.; Olsen, P.C.; Gazumyan, A.; Schaefer-Babajew, D.; Avila-Rios, S.; Nogueira, L.; Patel, R. Recurrent potent human neutralizing antibodies to Zika virus in Brazil and Mexico. *Cell* **2017**, *169*, 597–609. [CrossRef]
35. Kim, S.I.; Kim, S.; Shim, J.M.; Lee, H.J.; Chang, S.Y.; Park, S.; Min, J.Y.; Park, W.B.; Oh, M.D.; Kim, S.; et al. Neutralization of Zika virus by E protein domain III-Specific human monoclonal antibody. *Biochem. Biophys. Res. Commun.* **2021**, *545*, 33–39. [CrossRef]
36. Graham, S.D.; Tu, H.A.; McElvany, B.D.; Graham, N.R.; Grinyo, A.; Davidson, E.; Doranz, B.J.; Diehl, S.A.; de Silva, A.M.; Markmann, A.J. A Novel Antigenic Site Spanning Domains I and III of the Zika Virus Envelope Glycoprotein Is the Target of Strongly Neutralizing Human Monoclonal Antibodies. *J. Virol.* **2021**, *95*, e02423-20. [CrossRef]
37. Stettler, K.; Beltramello, M.; Espinosa, D.A.; Graham, V.; Cassotta, A.; Bianchi, S.; Vanzetta, F.; Minola, A.; Jaconi, S.; Mele, F. Specificity, cross-reactivity, and function of antibodies elicited by Zika virus infection. *Science* **2016**, *353*, 823–826. [CrossRef]
38. Dussupt, V.; Sankhala, R.S.; Gromowski, G.D.; Donofrio, G.; De La Barrera, R.A.; Larocca, R.A.; Zaky, W.; Mendez-Rivera, L.; Choe, M.; Davidson, E. Potent Zika and dengue cross-neutralizing antibodies induced by Zika vaccination in a dengue-experienced donor. *Nat. Med.* **2020**, *26*, 228–235. [CrossRef]
39. Barba-Spaeth, G.; Dejnirattisai, W.; Rouvinski, A.; Vaney, M.C.; Medits, I.; Sharma, A.; Simon-Loriere, E.; Sakuntabhai, A.; Cao-Lormeau, V.M.; Haouz, A.; et al. Structural basis of potent Zika-dengue virus antibody cross-neutralization. *Nature* **2016**, *536*, 48–53. [CrossRef]
40. Wang, J.; Bardelli, M.; Espinosa, D.A.; Pedotti, M.; Ng, T.S.; Bianchi, S.; Simonelli, L.; Lim, E.X.Y.; Foglierini, M.; Zatta, F.; et al. A Human Bi-specific Antibody against Zika Virus with High Therapeutic Potential. *Cell* **2017**, *171*, 229–241.e25. [CrossRef]
41. Deng, Y.-Q.; Dai, J.-X.; Ji, G.-H.; Jiang, T.; Wang, H.-J.; Yang, H.-o.; Tan, W.-L.; Liu, R.; Yu, M.; Ge, B.-X. A broadly flavivirus cross-neutralizing monoclonal antibody that recognizes a novel epitope within the fusion loop of E protein. *PLoS ONE* **2011**, *6*, e16059. [CrossRef] [PubMed]
42. França, R.; Silva, J.M.; Rodrigues, L.S.; Sokolowskei, D.; Brigido, M.M.; Maranhão, A.Q. New Anti-Flavivirus Fusion Loop Human Antibodies with Zika Virus-Neutralizing Potential. *Int. J. Mol. Sci.* **2022**, *23*, 7805. [CrossRef] [PubMed]
43. Qu, P.; Zhang, C.; Li, M.; Ma, W.; Xiong, P.; Liu, Q.; Zou, G.; Lavillette, D.; Yin, F.; Jin, X. A new class of broadly neutralizing antibodies that target the glycan loop of Zika virus envelope protein. *Cell Discov.* **2020**, *6*, 5. [CrossRef] [PubMed]
44. McLaury, A.R.; Haun, B.K.; To, A.; Mayerlen, L.; Medina, L.O.; Lai, C.Y.; Wong, T.A.S.; Nakano, E.; Strange, D.; Aquino, D.; et al. Characterization of Two Highly Specific Monoclonal Antibodies Targeting the Glycan Loop of the Zika Virus Envelope Protein. *Viral Immunol.* **2024**, *37*, 167–175. [CrossRef]

145. Wang, Q.; Yang, H.; Liu, X.; Dai, L.; Ma, T.; Qi, J.; Wong, G.; Peng, R.; Liu, S.; Li, J. Molecular determinants of human neutralizing antibodies isolated from a patient infected with Zika virus. *Sci. Transl. Med.* **2016**, *8*, 369ra179. [CrossRef]
146. Sankhala, R.S.; Dussupt, V.; Donofrio, G.; Gromowski, G.D.; Rafael, A.; Larocca, R.A.; Mendez-Rivera, L.; Lee, A.; Choe, M Zaky, W. Zika-specific neutralizing antibodies targeting inter-dimer envelope epitopes. *Cell Rep.* **2023**, *42*, 112942. [CrossRef]
147. Zhang, S.; Kostyuchenko, V.A.; Ng, T.S.; Lim, X.N.; Ooi, J.S.G.; Lambert, S.; Tan, T.Y.; Widman, D.G.; Shi, J.; Baric, R.S.; et a Neutralization mechanism of a highly potent antibody against Zika virus. *Nat. Commun.* **2016**, *7*, 13679. [CrossRef]
148. Fernandez, E.; Dejnirattisai, W.; Cao, B.; Scheaffer, S.M.; Supasa, P.; Wongwiwat, W.; Esakky, P.; Drury, A.; Mongkolsapaya, J Moley, K.H.; et al. Human antibodies to the dengue virus E-dimer epitope have therapeutic activity against Zika virus infectio *Nat. Immunol.* **2017**, *18*, 1261–1269. [CrossRef]
149. Sapparapu, G.; Fernandez, E.; Kose, N.; Bin, C.; Fox, J.M.; Bombardi, R.G.; Zhao, H.; Nelson, C.A.; Bryan, A.L.; Barnes, T.; et a Neutralizing human antibodies prevent Zika virus replication and fetal disease in mice. *Nature* **2016**, *540*, 443–447. [CrossRef]
150. Hasan, S.S.; Miller, A.; Sapparapu, G.; Fernandez, E.; Klose, T.; Long, F.; Fokine, A.; Porta, J.C.; Jiang, W.; Diamond, M.S. A huma antibody against Zika virus crosslinks the E protein to prevent infection. *Nat. Commun.* **2017**, *8*, 14722. [CrossRef]
151. Sariol, C.A.; Nogueira, M.L.; Vasilakis, N. A Tale of Two Viruses: Does Heterologous Flavivirus Immunity Enhance Zika Disease *Trends Microbiol.* **2018**, *26*, 186–190. [CrossRef] [PubMed]
152. Langerak, T.; Mumtaz, N.; Tolk, V.I.; van Gorp, E.C.M.; Martina, B.E.; Rockx, B.; Koopmans, M.P.G. The possible role c cross-reactive dengue virus antibodies in Zika virus pathogenesis. *PLoS Pathog.* **2019**, *15*, e1007640. [CrossRef] [PubMed]
153. Oliveira, R.A.S.; de Oliveira-Filho, E.F.; Fernandes Iv, A.; Brito, C.A.A.; Marques, E.T.A.; Tenório, M.C.; Gil, L.H.G.V. Previou dengue or Zika virus exposure can drive to infection enhancement or neutralisation of other flaviviruses. *Mem. Inst. Oswald Cruz* **2019**, *114*, e190098. [CrossRef] [PubMed]
154. Lessler, J.; Chaisson, L.H.; Kucirka, L.M.; Bi, Q.; Grantz, K.; Salje, H.; Carcelen, A.C.; Ott, C.T.; Sheffield, J.S.; Ferguson, N.M Assessing the global threat from Zika virus. *Science* **2016**, *353*, aaf8160. [CrossRef]
155. Castanha, P.M.S.; Marques, E.T.A. A glimmer of hope: Recent updates and future challenges in Zika vaccine development. *Virus* **2020**, *12*, 1371. [CrossRef]
156. Malafa, S.; Medits, I.; Aberle, J.H.; Aberle, S.W.; Haslwanter, D.; Tsouchnikas, G.; Wolfel, S.; Huber, K.L.; Percivalle, E.; Cherpillo P.; et al. Impact of flavivirus vaccine-induced immunity on primary Zika virus antibody response in humans. *PLoS Negl. Tro Dis.* **2020**, *14*, e0008034. [CrossRef]
157. Montecillo-Aguado, M.R.; Montes-Gomez, A.E.; Garcia-Cordero, J.; Corzo-Gomez, J.; Vivanco-Cid, H.; Mellado-Sanchez, C Munoz-Medina, J.E.; Gutierrez-Castaneda, B.; Santos-Argumedo, L.; Gonzalez-Bonilla, C.; et al. Cross-Reaction, Enhancemer and Neutralization Activity of Dengue Virus Antibodies against Zika Virus: A Study in the Mexican Population. *J. Immunol. R* **2019**, *2019*, 7239347. [CrossRef]
158. Wang, W.H.; Urbina, A.N.; Wu, C.C.; Lin, C.Y.; Thitithanyanont, A.; Assavalapsakul, W.; Lu, P.L.; Chen, Y.H.; Wang, S An epidemiological survey of the current status of Zika and the immune interaction between dengue and Zika infection Southern Taiwan. *Int. J. Infect. Dis.* **2020**, *93*, 151–159. [CrossRef]
159. Bonheur, A.N.; Thomas, S.; Soshnick, S.H.; McGibbon, E.; Dupuis, A.P., 2nd; Hull, R.; Slavinski, S.; Del Rosso, P.E.; Weiss, I Hunt, D.T.; et al. A fatal case report of antibody-dependent enhancement of dengue virus type 1 following remote Zika vir infection. *BMC Infect. Dis.* **2021**, *21*, 749. [CrossRef]
160. Bardina, S.V.; Bunduc, P.; Tripathi, S.; Duehr, J.; Frere, J.J.; Brown, J.A.; Nachbagauer, R.; Foster, G.A.; Krysztof, D.; Tortorella, I et al. Enhancement of Zika virus pathogenesis by preexisting antiflavivirus immunity. *Science* **2017**, *356*, 175–180. [CrossRef]
161. Crooks, C.M.; Weiler, A.M.; Rybarczyk, S.L.; Bliss, M.I.; Jaeger, A.S.; Murphy, M.E.; Simmons, H.A.; Mejia, A.; Fritsch, M.I Hayes, J.M.; et al. Previous exposure to dengue virus is associated with increased Zika virus burden at the maternal-fetal interfa in rhesus macaques. *PLoS Negl. Trop. Dis.* **2021**, *15*, e0009641. [CrossRef] [PubMed]
162. Shi, Y.; Dai, L.; Song, H.; Gao, G.F. Structures of Zika Virus E & NS1: Relations with Virus Infection and Host Immune Respons *Adv. Exp. Med. Biol.* **2018**, *1062*, 77–87. [CrossRef] [PubMed]
163. Wolford, R.W.; Schaefer, T.J. Zika Virus. In *StatPearls*; StatPearls Publishing: Treasure Island, FL, USA, 2024.
164. Figueiredo, L.T. The Brazilian flaviviruses. *Microbes Infect.* **2000**, *2*, 1643–1649. [CrossRef] [PubMed]
165. Anez, G.; Rios, M. Dengue in the United States of America: A worsening scenario? *BioMed Res. Int.* **2013**, *2013*, 678645. [CrossR
166. Chang, H.H.; Huber, R.G.; Bond, P.J.; Grad, Y.H.; Camerini, D.; Maurer-Stroh, S.; Lipsitch, M. Systematic analysis of prote identity between Zika virus and other arthropod-borne viruses. *Bull. World Health Organ.* **2017**, *95*, 517–525I. [CrossRef]
167. Sekaran, S.D.; Ismail, A.A.; Thergarajan, G.; Chandramathi, S.; Rahman, S.K.H.; Mani, R.R.; Jusof, F.F.; Lim, Y.A.L.; Manikam Host immune response against DENV and ZIKV infections. *Front. Cell. Infect. Microbiol.* **2022**, *12*, 975222. [CrossRef]
168. Dejnirattisai, W.; Supasa, P.; Wongwiwat, W.; Rouvinski, A.; Barba-Spaeth, G.; Duangchinda, T.; Sakuntabhai, A.; Cao-Lorme V.-M.; Malasit, P.; Rey, F.A.; et al. Dengue virus sero-cross-reactivity drives antibody-dependent enhancement of infection w zika virus. *Nat. Immunol.* **2016**, *17*, 1102–1108. [CrossRef]

169. Kribs, C.; Greenhalgh, D.J.J.o.M.B. Impact of tetravalent dengue vaccination with screening, ADE, and altered infectivity on single-serotype dengue and Zika transmission. *J. Math. Biol.* **2023**, *86*, 85. [CrossRef]
170. Xu, M.; Zuest, R.; Velumani, S.; Tukijan, F.; Toh, Y.X.; Appanna, R.; Tan, E.Y.; Cerny, D.; MacAry, P.; Wang, C.I.; et al. A potent neutralizing antibody with therapeutic potential against all four serotypes of dengue virus. *NPJ Vaccines* **2017**, *2*, 2. [CrossRef]
171. Kam, Y.-W.; Lee, C.Y.-P.; Teo, T.-H.; Howland, S.W.; Amrun, S.N.; Lum, F.-M.; See, P.; Kng, N.Q.-R.; Huber, R.G.; Xu, M.-H. Cross-reactive dengue human monoclonal antibody prevents severe pathologies and death from Zika virus infections. *JCI Insight* **2017**, *2*, e92428. [CrossRef]
172. Sun, H.; Yang, M.; Lai, H.; Neupane, B.; Teh, A.Y.; Jugler, C.; Ma, J.K.; Steinkellner, H.; Bai, F.; Chen, Q. A Dual-Approach Strategy to Optimize the Safety and Efficacy of Anti-Zika Virus Monoclonal Antibody Therapeutics. *Viruses* **2023**, *15*, 1156. [CrossRef] [PubMed]
173. Khandia, R.; Munjal, A.; Dhama, K.; Karthik, K.; Tiwari, R.; Malik, Y.S.; Singh, R.K.; Chaicumpa, W. Modulation of Dengue/Zika Virus Pathogenicity by Antibody-Dependent Enhancement and Strategies to Protect Against Enhancement in Zika Virus Infection. *Front. Immunol.* **2018**, *9*, 597. [CrossRef] [PubMed]
174. Dowd, K.A.; Pierson, T.C. Antibody-mediated neutralization of flaviviruses: A reductionist view. *Virology* **2011**, *411*, 306–315. [CrossRef] [PubMed]
175. Sharma, N.; Prosser, O.; Kumar, P.; Tuplin, A.; Giri, R. Small molecule inhibitors possibly targeting the rearrangement of Zika virus envelope protein. *Antivir. Res.* **2020**, *182*, 104876. [CrossRef]
176. Ci, Y.; Yao, B.; Yue, K.; Yang, Y.; Xu, C.; Li, D.F.; Qin, C.F.; Shi, L. Bortezomib inhibits ZIKV/DENV by interfering with viral polyprotein cleavage via the ERAD pathway. *Cell Chem. Biol.* **2023**, *30*, 527–539.e5. [CrossRef]
177. Gao, Y.; Tai, W.; Wang, N.; Li, X.; Jiang, S.; Debnath, A.K.; Du, L.; Chen, S. Identification of Novel Natural Products as Effective and Broad-Spectrum Anti-Zika Virus Inhibitors. *Viruses* **2019**, *11*, 1019. [CrossRef]
178. Carneiro, B.M.; Batista, M.N.; Braga, A.C.S.; Nogueira, M.L.; Rahal, P. The green tea molecule EGCG inhibits Zika virus entry. *Virology* **2016**, *496*, 215–218. [CrossRef]
179. Behrendt, P.; Perin, P.; Menzel, N.; Banda, D.; Pfaender, S.; Alves, M.P.; Thiel, V.; Meuleman, P.; Colpitts, C.C.; Schang, L.M. Pentagalloylglucose, a highly bioavailable polyphenolic compound present in Cortex moutan, efficiently blocks hepatitis C virus entry. *Antivir. Res.* **2017**, *147*, 19–28. [CrossRef]
180. Sharma, N.; Kumar, P.; Giri, R. Polysaccharides like pentagalloylglucose, parishin a and stevioside inhibits the viral entry by binding the Zika virus envelope protein. *J. Biomol. Struct. Dyn.* **2021**, *39*, 6008–6020. [CrossRef]
181. Sangeetha, K.; Martin-Acebes, M.A.; Saiz, J.C.; Meena, K.S. Molecular docking and antiviral activities of plant derived compounds against zika virus. *Microb. Pathog.* **2020**, *149*, 104540. [CrossRef]
182. Oo, A.; Teoh, B.T.; Sam, S.S.; Bakar, S.A.; Zandi, K. Baicalein and baicalin as Zika virus inhibitors. *Arch. Virol.* **2019**, *164*, 585–593. [CrossRef] [PubMed]
183. Ho, Y.-J.; Lu, J.-W.; Huang, Y.-L.; Lai, Z.-Z. Palmatine inhibits Zika virus infection by disrupting virus binding, entry, and stability. *Biochem. Biophys. Res. Commun.* **2019**, *518*, 732–738. [CrossRef] [PubMed]
184. Lai, Z.-Z.; Ho, Y.-J.; Lu, J.-W. Harringtonine inhibits Zika virus infection through multiple mechanisms. *Molecules* **2020**, *25*, 4082. [CrossRef] [PubMed]
185. Liu, S.; DeLalio, L.J.; Isakson, B.E.; Wang, T.T. AXL-Mediated Productive Infection of Human Endothelial Cells by Zika Virus. *Circ. Res.* **2016**, *119*, 1183–1189. [CrossRef]
186. Zhang, B.; Yu, J.; Zhu, G.; Huang, Y.; Zhang, K.; Xiao, X.; He, W.; Yuan, J.; Gao, X. Dapoxetine, a Selective Serotonin Reuptake Inhibitor, Suppresses Zika Virus Infection In Vitro. *Molecules* **2023**, *28*, 8142. [CrossRef]
187. Fernando, S.; Fernando, T.; Stefanik, M.; Eyer, L.; Ruzek, D. An Approach for Zika Virus Inhibition Using Homology Structure of the Envelope Protein. *Mol. Biotechnol.* **2016**, *58*, 801–806. [CrossRef]
188. Mounce, B.C.; Cesaro, T.; Carrau, L.; Vallet, T.; Vignuzzi, M. Curcumin inhibits Zika and chikungunya virus infection by inhibiting cell binding. *Antivir. Res.* **2017**, *142*, 148–157. [CrossRef]
189. Kim, M.; Choi, H.; Kim, Y.B. Therapeutic targets and biological mechanisms of action of curcumin against Zika virus: In silico and in vitro analyses. *Eur. J. Pharmacol.* **2021**, *904*, 174144. [CrossRef]
190. Pitts, J.; Hsia, C.Y.; Lian, W.; Wang, J.; Pfeil, M.P.; Kwiatkowski, N.; Li, Z.; Jang, J.; Gray, N.S.; Yang, P.L. Identification of small molecule inhibitors targeting the Zika virus envelope protein. *Antivir. Res.* **2019**, *164*, 147–153. [CrossRef]
191. Li, F.; Lang, Y.; Ji, Z.; Xia, Z.; Han, Y.; Cheng, Y.; Liu, G.; Sun, F.; Zhao, Y.; Gao, M. A scorpion venom peptide Ev37 restricts viral late entry by alkalizing acidic organelles. *J. Biol. Chem.* **2019**, *294*, 182–194. [CrossRef]
192. Clark, H.L.; Minns, M.S.; Sun, Y.; de Jesus, T.; Ghannoum, M.G.; Pearlman, E. Atovaquone impairs growth of Aspergillus and Fusarium keratitis isolates by modulating mitochondrial function and zinc homeostasis. *Investig. Ophthalmol. Vis. Sci.* **2018**, *59*, 1589–1598. [CrossRef] [PubMed]
193. Nixon, G.L.; Moss, D.M.; Shone, A.E.; Lalloo, D.G.; Fisher, N.; O'Neill, P.M.; Ward, S.A.; Biagini, G.A. Antimalarial pharmacology and therapeutics of atovaquone. *J. Antimicrob. Chemother.* **2013**, *68*, 977–985. [CrossRef] [PubMed]

194. Yamamoto, M.; Ichinohe, T.; Watanabe, A.; Kobayashi, A.; Zhang, R.; Song, J.; Kawaguchi, Y.; Matsuda, Z.; Inoue, J.-i. The antimalarial compound atovaquone inhibits Zika and dengue virus infection by blocking E protein-mediated membrane fusion. *Viruses* **2020**, *12*, 1475. [CrossRef] [PubMed]
195. Wang, L.; Liang, R.; Gao, Y.; Li, Y.; Deng, X.; Xiang, R.; Zhang, Y.; Ying, T.; Jiang, S.; Yu, F. Development of Small-Molecule Inhibitors Against Zika Virus Infection. *Front. Microbiol.* **2019**, *10*, 2725. [CrossRef]
196. Yu, Y.; Deng, Y.Q.; Zou, P.; Wang, Q.; Dai, Y.; Yu, F.; Du, L.; Zhang, N.N.; Tian, M.; Hao, J.N.; et al. A peptide-based viral inactivator inhibits Zika virus infection in pregnant mice and fetuses. *Nat. Commun.* **2017**, *8*, 15672. [CrossRef]
197. Chen, L.; Liu, Y.; Wang, S.; Sun, J.; Wang, P.; Xin, Q.; Zhang, L.; Xiao, G.; Wang, W. Antiviral activity of peptide inhibitors derived from the protein E stem against Japanese encephalitis and Zika viruses. *Antivir. Res.* **2017**, *141*, 140–149. [CrossRef]
198. Hu, D.; Wang, Y.; Li, A.; Li, Q.; Wu, C.; Shereen, M.A.; Huang, S.; Wu, K.; Zhu, Y.; Wang, W. LAMR1 restricts Zika virus infection by attenuating the envelope protein ubiquitination. *Virulence* **2021**, *12*, 1795–1807. [CrossRef]
199. Ling, J.; Morente, S.F.; Lundkvist, Å.; Li, J. Heparanase attenuates Zika virus infection by destabilizing the viral envelope protein *bioRxiv* **2025**. [CrossRef]
200. Van der Hoek, K.H.; Eyre, N.S.; Shue, B.; Khantisitthiporn, O.; Glab-Ampi, K.; Carr, J.M.; Gartner, M.J.; Jolly, L.A.; Thomas, P.Q. Adikusuma, F.; et al. Viperin is an important host restriction factor in control of Zika virus infection. *Sci. Rep.* **2017**, *7*, 4475 [CrossRef]
201. Wang, Y.; Li, Q.; Hu, D.; Gao, D.; Wang, W.; Wu, K.; Wu, J. USP38 Inhibits Zika Virus Infection by Removing Envelope Protein Ubiquitination. *Viruses* **2021**, *13*, 2099. [CrossRef]
202. Shan, C.; Xie, X.; Muruato, A.E.; Rossi, S.L.; Roundy, C.M.; Azar, S.R.; Yang, Y.; Tesh, R.B.; Bourne, N.; Barrett, A.D. An infectious cDNA clone of Zika virus to study viral virulence, mosquito transmission, and antiviral inhibitors. *Cell Host Microbe* **2016** *19*, 891–900. [CrossRef] [PubMed]
203. Annamalai, A.S.; Pattnaik, A.; Sahoo, B.R.; Guinn, Z.P.; Bullard, B.L.; Weaver, E.A.; Steffen, D.; Natarajan, S.K.; Petro, T.M. Pattnaik, A.K. An Attenuated Zika Virus Encoding Non-Glycosylated Envelope (E) and Non-Structural Protein 1 (NS1) Confers Complete Protection against Lethal Challenge in a Mouse Model. *Vaccines* **2019**, *7*, 112. [CrossRef] [PubMed]
204. Zhong, Z.; Portela Catani, J.P.; Mc Cafferty, S.; Couck, L.; Van Den Broeck, W.; Gorlé, N.; Vandenbroucke, R.E.; Devriendt, B. Ulbert, S.; Cnops, L. Immunogenicity and protection efficacy of a naked self-replicating mRNA-based Zika virus vaccine. *Vaccines* **2019**, *7*, 96. [CrossRef] [PubMed]
205. Contreras, M.; Stuart, J.B.; Levoir, L.M.; Belmont, L.; Goo, L. Defining the impact of flavivirus envelope protein glycosylation site mutations on sensitivity to broadly neutralizing antibodies. *mBio* **2024**, *15*, e03048-23. [CrossRef]
206. Feng, Y.J.F.i.P. Recent advances in the study of zika virus structure, drug targets, and inhibitors. *Front. Pharmacol.* **2024** *15*, 1418516. [CrossRef]
207. Khan, E.; Jindal, H.; Mishra, P.; Suvvari, T.K.; Jonna, S. The 2021 Zika outbreak in Uttar Pradesh state of India: Tackling the emerging public health threat. *Trop. Dr.* **2022**, *52*, 474–478. [CrossRef]
208. Regla-Nava, J.A.; Wang, Y.T.; Fontes-Garfias, C.R.; Liu, Y.; Syed, T.; Susantono, M.; Gonzalez, A.; Viramontes, K.M.; Verma, S.K Kim, K.; et al. A Zika virus mutation enhances transmission potential and confers escape from protective dengue virus immunity *Cell Rep.* **2022**, *39*, 110655. [CrossRef]
209. Chan, K.W.K.; Bifani, A.M.; Watanabe, S.; Choy, M.M.; Ooi, E.E.; Vasudevan, S.G. Tissue-specific expansion of Zika virus isogenic variants drive disease pathogenesis. *eBioMedicine* **2023**, *91*, 104570. [CrossRef]
210. Giraldo, M.I.; Gonzalez-Orozco, M.; Rajsbaum, R. Pathogenesis of Zika Virus Infection. *Annu. Rev. Pathol.* **2023**, *18*, 181–20. [CrossRef]

viruses

MDPI

Perspective

Building Local Research Capacity for Global Pandemic Preparedness: Lessons from WHO Unity Studies and Their Expansion in India

Yasir Alvi [1,*], Farzana Islam [1,*], Mohammad Ahmad [2], Richa Gautam [1], Aqsa Shaikh [1], Musharraf Husain [3], Kartikey Yadav [1], Mohammad Rashid [1], Shyambhavee Behera [1], Nicki L. Boddington [4], Ashok Basnet [5] and Pushpa Ranjan Wijesinghe [5]

1 Department of Community Medicine, Hamdard Institute of Medical Sciences and Research, New Delhi 110062, India; richa.gautam1@gmail.com (R.G.); emailtoaqsa@gmail.com (A.S.); drkartikeyy@gmail.com (K.Y.); mohammad.rashid465@gmail.com (M.R.); shyambhavee@gmail.com (S.B.)
2 World Health Organization (WHO), New Delhi 110062, India
3 Department of Surgery, Hamdard Institute of Medical Sciences and Research, New Delhi 110062, India
4 WHO Health Emergencies Programme, World Health Organization, CH-1211 Geneva, Switzerland
5 WHO Health Emergencies Programme, South-East Asia Regional Office, New Delhi 110002, India
* Correspondence: yasiralvi13@gmail.com (Y.A.); drfarzanaislam@gmail.com (F.I.)

Abstract

The rapid onset and progression of the COVID-19 pandemic highlighted the critical necessity for standardized, timely epidemiological investigations to generate actionable evidence for public health policy. The World Health Organization (WHO) Global Influenza Surveillance and Response System (GISRS) and Unity Studies and Investigations initiative (Unity Studies) provides the standardized framework to address these critical knowledge gaps. This manuscript reflects upon the Hamdard Institute of Medical Sciences and Research (HIMSR)'s experience as an active site implementing three WHO Unity protocols between 2020 and 2021. We synthesize key findings from the Household Transmission Investigation (HHTI) and the Health Facility Transmission (HCW cohort) studies, detail the operational and analytical complexities addressed through intensive collaboration with WHO HQ, South-East Asia Regional Office (SEARO), and WHO India, and outline the subsequent institutional capacity transmission. Building directly on this established expertise, HIMSR has been designated as the dedicated Nodal network site for the WHO SEAR Unity Studies Network in India, coordinating administration activities of a vast network of national institutes for ongoing pandemic preparedness. This trajectory demonstrates the potential for low- and middle-income country (LMIC) institutions not only to contribute critical evidence during crises but also to transition into resilient national and regional research and surveillance platforms for future pan-respiratory pathogen threats. We detail the essential findings and operational lessons from the COVID-19 pandemic response and elaborate extensively on the strategic implementation plan for the proposed WHO Unity Nodal Network site in India, emphasizing capacity building, standardization, and the integration of research into public health policy.

Keywords: pandemic preparedness; Unity studies; Influenza; disease transmission; epidemic; public health action; infection prevention and control

1. Introduction

The advent of Severe Acute Respiratory Syndrome Coronavirus 2 (SARS-CoV-2) initiated the COVID-19 pandemic, creating uncertainty over the virus's epidemiology and transmissibility properties [1]. To quickly gather evidence-based knowledge and guide public health actions, the World Health Organization (WHO) launched the Investigations and Studies Global Initiative, also known as the Unity Studies [2]. The Unity Studies offer a generic framework for preparedness and readiness for responding to epidemics and pandemics, rapidly adapted from existing protocols (including the Consortium for the Standardization of Influenza Seroepidemiology (CONSISE)), for carrying out standardized investigations essential for the risk assessment of emerging respiratory pathogens [2–4]. This standardized architecture guarantees systematic data collection and promotes international comparability, allowing the global community to collaboratively address knowledge gaps [5,6].

Standardized protocols, including the Household Transmission Investigation (HHTI) and the Healthcare Worker (HCW) risk assessment, were crucial for addressing key uncertainties related to transmissibility, severity, and the extent of asymptomatic infection [2–4]. This was conducted alongside giving priority to research equity, where these studies supported LMIC serological tests and technical support [2].

The Hamdard Institute of Medical Sciences and Research (HIMSR), New Delhi, is a medical college established in 2011. It provides quality healthcare at an affordable cost supported by the 710 bed Hakeem Abdul Hameed Centenary (HAHC) Hospital, and is well equipped with National Accreditation Board for Testing and Calibration Laboratories (NABL) having adequate biosafety level. A separate Central Research lab is an added feather in the crown of HIMSR. It was one of the sites of WHO's Unity studies global network during this important time. Between December 2020 and July 2021, the team successfully adapted and completed three different WHO Unity protocols and published six studies in high-index journals [1,7–11]. These were as follows:

1. Household Transmission Investigation (HHTI): A prospective case-ascertained study of COVID-19 transmission dynamics [1,8,10].
2. Health Care Worker (HCW) Cohort Study: An assessment of infection risk factors and prevention measures among healthcare personnel [7,11].
3. COVID-19 Vaccine Effectiveness (VE) Study: A case-control study conducted in collaboration with the Indian Council of Medical Research (ICMR) and WHO [9].

These studies offered timely local data from Delhi, which were helpful during the second wave of COVID-19, possibly driven by the B.1.617.2 "Delta" variant. As a result of the successful implementation of these standardized protocols, coupled with working closely with district and state authorities and WHO offices (HQ, SEARO, and India Country Office) and subsequent academic publications, HIMSR became a significant institution for contributing in generating evidence for national and global pandemic planning. This manuscript discusses the lessons learned from these projects and details HIMSR's ongoing transition into Nodal network site of the WHO Global Influenza Surveillance and Response System (GISRS) and Unity Studies and Investigations initiative (WHO Unity 2.0) [2,12]. We also tried to describe the programmatic data concerning the operational challenges, capacity strides made, and compilation of the experiences from all three studies at a programmatic level. The goal is to build preparedness, response and resilience against future respiratory threats, in line with the Preparedness and Resilience for Emerging Threats (PRET) framework and the recently accepted WHO Pandemic Agreement [2,12,13].

https://doi.org/10.3390/v18020

2. The Rationale for Focused Unity Network in India

2.1. Indian Context

The Indian context poses complex public health challenges, characterized by high population density, which were significantly exacerbated during the COVID-19 pandemic and its prevention activities. In early 2021, India had the second-highest number of COVID-19 cases in the globe, and New Delhi saw some of the worst spikes [1]. Overcrowding and poor housing conditions can make lockdowns and isolation less effective, which could make it more likely for close contact to spread the virus. The HIMSR HHTI study focused specifically on these dynamics in Delhi's South and South-East districts, leading to the significant discovery of a high secondary infection rate (SIR) of 44.6% among unvaccinated household contacts, which was much higher than the average for the whole world [1]. The local data also found risk factors that were specific to the situation, such as being female and taking care of the index case. This means that women, who often act as primary caretakers at the household level, had a higher risk. A distinctive finding in the local context was that an increased number of rooms was substantially correlated with secondary infection. This acute demand for clear local epidemiological data happened at the same time as a systemic need in India to improve the ability to gather evidence that is useful for health policies [1,7].

2.2. Philosophy and Origin

The operational philosophy of HIMSR's participation in the Unity Studies initiative is based on the goal of promoting research equity and strengthening pandemic preparedness through standardized operational research [2]. Building on the GISRS platform—which promotes effective collaboration, virus data sharing with the WHO for global risk assessment, and their commitment to a global public health framework for rapid evidence generation—the Unity Studies initiative emerged from the recognized need after the 2009 H1N1 pandemic to standardize methodologies for early investigations of emerging respiratory pathogens [2,6]. This led to a generic preparedness and readiness framework for influenza pandemics, which was rapidly adapted for SARS-CoV-2 in January 2020 [5].

For HIMSR, implementing the Unity protocols meant moving from passive surveillance to systematically investigating transmission dynamics and the whole spectrum of disease, including asymptomatic infection, within closed populations. By adopting WHO's Unity Studies protocols, the HIMSR team was able to use standardized methods, obtain technical support and funding (including from the WHO Unity Studies program), and use data management system tools, thus reinforcing local research resilience [3,4]. This standardization ensured that the data collected were robust and internationally comparable and also could be used for triangulation and providing global and regional estimates [2,10]. This alignment directly supported the strategic aim of building capacity in LMICs. The strong institutional network was cemented through interaction and guidance from WHO Country Office staff and WHO SEARO, positioning HIMSR as a regional center of expertise [13]. These types of strategic partnership between the WHO and research institutions provided a platform which ensures that future research activities are meticulously organized and primed for swift implementation during future health emergencies, maintaining focus on LMIC capacity building as well as contributing to global risk assessments [2,13].

3. Overview of the WHO Unity Platform

The WHO GISRS Unity Studies and Investigations initiative, or Unity Studies, is an innovative way of integrating surveillance techniques and systemic research. This initiative offers a robust global platform for pandemic preparedness, targeting emerging or re-emerging respiratory pathogens [2,13,14]. Unity Studies serve as essential instruments

https://doi.org/10.3390/v18020198

that supplement routine surveillance systems by addressing specific public health inquiries concerning transmissibility, population immunity, and infection/disease severity that routine systems may not efficiently capture [2]. The framework ensures data gathered across diverse settings is comparable and robust for pooled analysis [2,6]. The core of the platform is the collection of standardized template protocols, developed in collaboration with technical partners like CONSISE and the University of Melbourne [2]. Some of the most important sorts of investigations are as follows:

- First Few X (FFX) Cases and Contacts Protocol: For early and rapid data collection on initial cases.
- Household Transmission Investigation (HHTI): A prospective case-ascertained study focusing on transmission in defined household settings.
- Health Worker (HCW) Risk Factor Protocol: A cohort study examining transmission and the effectiveness of Infection Prevention and Control (IPC) measures in healthcare settings.
- Population-based age-stratified sero-epidemiological investigation protocol: Used to measure population immunity/seroprevalence.
- Vaccine Effectiveness (VE) Protocols: For rapid assessment of the effectiveness of vaccines.

These protocols let researchers estimate multiple epidemiological parameters across the disease pyramid, from asymptomatic infection at the base to disease severity and fatality ratios at the top (Figure 1).

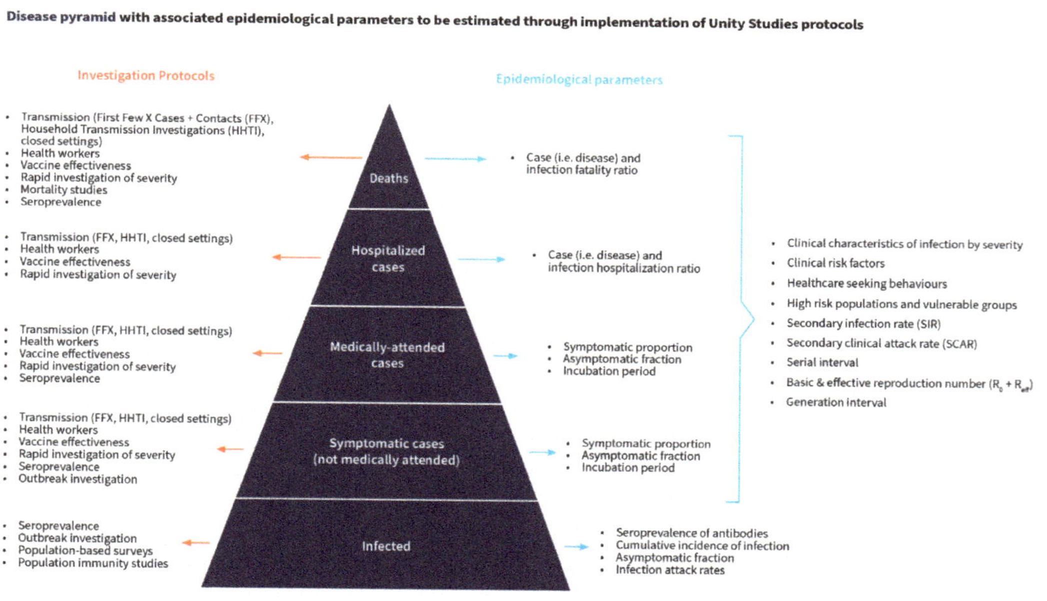

Figure 1. Disease pyramid with associated epidemiological parameters to be estimated through implementation of Unity Studies protocol, Adapted with permission Bergeri et al., 2024 [2,15].

The current Unity Studies 2.0 focuses on sustaining operational readiness through the development of a Global Operational Network of Sites. This network is aligned with the PRET initiative, ensuring readiness is exercised during "peace time" respiratory outbreaks, such as seasonal Influenza and Respiratory Syncytial Virus (RSV) epidemics. HIMSR has

https://doi.org/10.3390/v180201

formalized its commitment to this network by signing the Terms of Reference (ToR) in December 2024 [13]

4. Reflections on Program Design and Implementation

HIMSR successfully implemented and completed three Unity projects during 2020–2021, covering the periods when India's first wave of COVID-19 was receding and the second wave was peaking.

The first project was HHTI [1]. This prospective, case-ascertained study used the WHO Unity HHTI protocol (Version 2.2) [3]. The primary objective was to learn about transmission dynamics and identify associated risk factors of secondary infection among households' contacts (HHC) of COVID-19 in South and South-East Delhi. The study enrolled 99 index cases and 316 HHCs. The study employed a four-dimensional conceptual framework to examine factors affecting transmission: index case characteristics, household level factors, individual level factors, and contact patterns (Figure 2). Data, along with biological samples (RT-PCR and paired serum samples for serology using WANTAI kits), were collected during four household visits on days 1, 7, 14, and 28 [1].

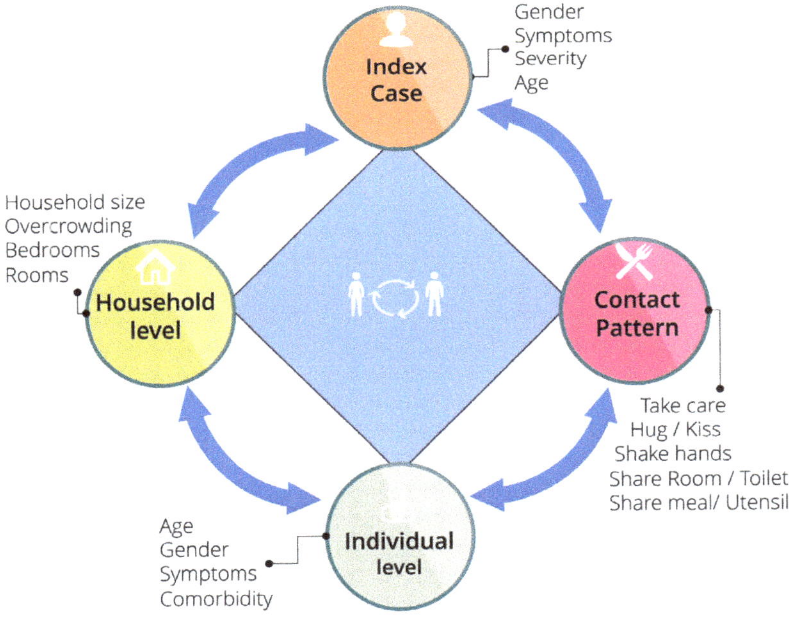

Figure 2. Factors governing transmission dynamics among household members across four dimensions. The transmission of disease among household members is affected by the interaction of four sets of factors, including index case characteristics (age, gender, and symptoms and their severity), household level (household size, number of bedrooms, room for isolation, overcrowding), individual level (age, gender, symptoms, comorbidity), and contact pattern (taking care, hugging, kissing, handshaking, sharing room, toilet, meals, and utensils); Adapted with permission from Islam et al., 2024 [1].

The key results were (HHTI) as follows:

- High Transmission: The secondary infection rate (SIR) was exceptionally high at 44.6% (95% CI 39.1–50.1). Cumulative secondary attack rate (SAR) was 55.5%, and the household SAR was 54.9%.

https://doi.org/10.3390/v18020198

- Epidemiological Parameters: Mean serial interval was 7.97 days (±6.66), and the effective reproduction number 1.33 (±1.56) and detectable shedding was 12.49 days (index cases) and 11.84 days (contacts).
- The four dimensions transmission dynamics showed significant associations with secondary infection:
    - Individual Level: Being female (OR 2.13), increasing age (OR 1.01), and symptoms at baseline (OR 3.39) or during follow-up (OR 3.18).
    - Index Case Level: Symptoms during follow-up (OR 6.29) and increasing age (OR 1.03).
    - Household Level: Having more rooms (OR 4.44).
    - Contact Pattern: Taking care of the index case (OR 2.02).

The second project was the Health Care Worker (HCW) Prospective Cohort Study [7]. This study adhered to the WHO Unity protocol for risk assessment among HCWs, identifying risk of disease transmission among those who were exposed to COVID-19 patients at HIMSR and HAHC Hospital [4]. The study was conducted from December 2020 to June 2021 and enrolled 192 HCWs, overcoming initial exclusion criteria against seropositive individuals through a protocol deviation implemented after the national vaccination campaign launched on January 16, 2021. Serology (using WANTAI kits) was collected at baseline and endline (28 days). We found a high (62%) baseline seropositivity, which rose to 77.7% at endline. We also document the Seroconversion Rate at 36.7% overall. Doctors showed the highest rate (63.2%), followed by nurses (42.9%) and paramedics (13.0%). Vaccination has a major risk status for seropositivity at baseline (partial: OR 3.303; full: OR 2.428) and seroconversion (fully vaccinated OR 32.63). High adherence to IPC was generally reported (e.g., 94.7% of nurses used gowns in close contact). However, the study did not find a statistically significant association between adherence to IPC measures (including PPE use and hand hygiene) and seroconversion/titer increase. This finding was likely confounded by the high vaccination rates.

The third project was the Vaccine Effectiveness (VE) case control study [9]. It was a hospital-based, multicentric study with severe COVID-19 patients as the cases and those with COVID-19 negative as the control. The random effects logistic regression model to calculate the adjusted odds ratios (aOR) with a 95% confidence interval (CI) after adjusting for relevant known confounders was used. A total of 1143 cases and 2541 control patients were enrolled across 11 hospitals in India. The VE of complete vaccination was 85% (95% CI: 79–89%) with AZD1222/Covishield and 71% (95% CI: 57–81%) with BBV152/Covaxin [9].

Challenges of implementing Unity **Studies at HIMSR:** The study faced operational challenges, which are typical of emergency research. As these protocols were based on global generic Unity protocols and needed to be adopted and implemented at quick pace at the early phase of the COVID-19 pandemic, shortcomings existed. These gaps were later identified during the implementation phase and protocol deviations were necessary. Further, due to variable transmission waves, there were times of declining cases and their subsequent study enrollment, an retrospective enrolment strategies were implemented and added as protocol deviation. Calculating precise parameters like the incubation period (IP), serial interval, and Basic Reproduction Number was complicated by continuous daily household interaction, making the exact date of exposure difficult to ascertain. Furthermore, analyzing co-primary cases increased the complexity of estimating secondary infection rates. In the HCW study, we faced the issue of high baseline non-response and significant loss to follow-up. The unpredicted and early vaccination campaign for HCW introduced significant confounding variables, preventing clear immunological differentiation between infection-induced and vaccine-induced seroconversion. The VE study at HIMSR too faced severe shortages of case recruitment during the study period. As a measure, the shortage

https://doi.org/10.3390/v18020

was compensated by additional participant recruitments at other study sites within this multicentric study.

Programmatic Strengths: The projects benefited immensely from strong WHO support for protocol finalization, financial oversight, monitoring, analysis, and dissemination. State health administration and district health authorities too extended technical support and access to district contact tracing data for comprehensive data management and contact tracing, which were some of the major pillars of successful implementation.

5. Reflections on the Strides Made and Challenges Faced by the Program

The three COVID-19 Unity Studies conducted by the HIMSR team had both significant achievements and inherent difficulties in epidemiological research. The study team held multiple brainstorming and problem-solving sessions during and after the studies to understand the underlying causes of these challenges and to identify potential solutions, all of which are discussed below.

Strides Made (Individual and Collective Capacity)

1. *Generation of High-Quality, Contextual Evidence:* HIMSR successfully delivered a comprehensive understanding of the COVID-19 epidemic specific to Delhi, India during a period of high transmission. This evidence included important estimates such as the high Household SAR (54.9%) and risk factors (e.g., caregiving, gender, dwelling space). This strict adherence to standard methods made it possible for local knowledge to be shared and collaborated with global networks.
2. *Strengthening the Validity of Unity Protocols in LMICs:* The success validated the operational utility of complex, multi-visit, longitudinal WHO Unity protocols (HHTI, HCW) in India, which included sophisticated data collection and sequential biological sampling (RT-PCR and serology).
3. *Advanced Technical and Analytical Capacity:* The WHO trainings provided the team the opportunity to learn sophisticated skills in planning and analysis, including calculating complex parameters.
4. *Established Networking and Global Visibility:* Strong operational partnerships were built with WHO and with national institutes. This infrastructure, facilitated by key WHO personnel, positioned HIMSR as a regional center of expertise and contributed to global standardization efforts.
5. *Multidisciplinary Workforce Development:* The projects required the recruitment and training of specialized staff (trained healthcare workers, phlebotomists, data managers), enhancing local capacity for rapid deployment research.

Challenges Faced

1. *Methodological Ambiguity in Key Parameter Calculation:* As discussed earlier, it was hard to obtain accurate estimates of parameters like the incubation period and the serial interval due to continuous daily household exposure and the non-specificity of symptomatic definitions.
2. *Sampling and Follow-up Biases:* The use of convenience sampling (in HHTI and HCW cohort studies) for index case recruitment and the high rates of non-response and loss to follow-up may have led to possible selection bias, validity, questionable generalizability and lower statistical accuracy. The exclusion of hospitalized participation in the HHTI study resulted in the cohort omitting the most severely affected segment of the population, but was essential exclusion criteria as per HHTI Unity protocol.
3. *Low enrolments during inter-peak transmission:* Our studies (HHTI and HCW cohort studies) spanned through many transmission cycles, with very high infection transmission in the early phase and with new strain and dropping to a low level in between.

https://doi.org/10.3390/v18020198

With low positivity during inter-peak transmission combined with high rates of non-response, we had to adapt protocol deviation and enrolled participants retrospectively. In HHTI, to help with enrolment, we also expanded our target population, involved other districts in Delhi, and worked with their district health authorities for data sharing. The VE trial study also faced low enrolment and other partner sites stepped in to compensate the low enrolment at HIMSR.

4. *Confounding by Vaccination and IPC Measures:* The unexpected early rollout of the mass HCW vaccination campaign in the middle of the study severely confounded the results, especially in the HCW cohort. It became impossible to immunologically differentiate between infection-induced versus vaccine-elicited immune responses. Furthermore, despite reporting high adherence to PPE and robust institutional IPC measures, in the HCW study, we failed to find a significant protective association between individual IPC measures and reduced seroconversion risk.
5. *Resource and Logistics Constraints:* Administrative delays impacted timelines. The need to adapt protocols under time, budget, and human resource constraints led to limitations in ancillary objectives, such as deferring genotyping studies, which would have been useful for understanding variant transmission.

6. Proposed Unity Network for HHTI in India

The foundational capacity built by HIMSR during the COVID-19 pandemic response facilitated its transition and positioned it as a regional network site within the WHO Unity Studies Network (WHO Unity 2.0) [15]. This strategic move formalizes HIMSR's commitment to resilient pandemic preparedness, aligning with the global goals of the PRET and WHOs Mosaic respiratory surveillance framework [2]. This initiative is centered on exercising standardized protocols during inter-pandemic "peace time" with quick action during novel disease outbreaks [2,16]. Below, we have detailed the network's plans for structure, methodology, network expansion, and capacity building, emphasizing the transition from local expertise to a national resource.

A. Strategic Alignment and Core Objectives of the Unity Studies network site at HIMSR:

The HIMSR network site is strategically located in the WHO South-East Asia Regional Office (SEARO) chapter of the Unity Studies network, where it is responsible for carrying out the standardized Unity Studies. This network aims to increase the evidence-based knowledge for action, promote international comparability of epidemiological investigations, and support national public health and social measures [15]. The Unity Studies 2.0 initiative requires the establishment of a global operational network of sites prepared for the rapid assessment of novel respiratory viruses [2]. HIMSR solidified this commitment by having its senior management sign the Terms of Reference (ToR) with the WHO in December 2024 [13].

HIMSR's role as a regional network site is primary in systematically implementing the rigorous methodology and exercising investigations at Nodal and satellite sites during inter-pandemics (e.g., seasonal influenza) to maintain operational readiness [2,15]. This aligns with the "Mosaic Respiratory Surveillance Framework," where Unity Studies act as specialized tools that enhance routine systems [2,14]. In the proposed network, HIMSR will function as a Unity Studies network site, responsible for liaising with the WHO, coordinating administration activities, conducting training, consolidating reporting, and carrying out the mandatory tasks of adapting and exercising one of the Unity protocols during the inter-pandemic period. This is being piloted by HIMSR's adaptation of HHTI in an ongoing study [16]. This "exercising" phase is crucial for strengthening preparedness and rapid response mechanisms. This comprehensive approach ensures that HIMSR generates essential new knowledge and integrates existing fragmented regional evidence to guide policy, vaccination strategies, and public health capacity thus ensuring preparedness for a future pandemic [13,17].

B. Exercising Readiness—Primary Research Focus for 2025: HHTI on Influenza and RSV

The immediate priority is exercising the core HHTI protocol adapted for Influenza A, Influenza B, and Respiratory Syncytial Virus (RSV) among Delhi residents, commencing in 2025 [13,16,17]. This is justified by the rising burden of these viruses and aligns with WHO guidance and HIMSR's proven institutional expertise in respiratory pathogen research [17]. Given the deteriorating air quality in India's capital, this research may also uncover the extent of undocumented respiratory illnesses [18].

C. Protocol Adaptation and Scope Limitation: Based on lessons of resource constraints during the COVID-19 projects, the ongoing HHTI study is designated as a pilot study with strict scope limitations [17]:

- Pathogen: Focus shifted from a novel pathogen to Influenza A and B and RSV. Testing utilizes RT-PCR via the GeneXpert system (Xpert Xpress Flu/RSV test cartridges) for the simultaneous detection and differentiation of Influenza A and B and RSV viral RNA, utilizing in-house accredited PCR facilities
- Sampling Strategy (Pilot Phase—Oct to Nov 2025): The initial pilot is strictly constrained by time, budget, and human resources. It targets a small sample size of five index cases and all their household contacts to validate the adapted methodology before scaling.
- Data Collection Schedule: Simplified sampling schedule with only nasal samples on day 1 and day 7, with a symptom diary kept until day 14. Serological testing is not planned for this phase.
- Deferred Objectives: Advanced analyses, including the calculation of incubation period, serial interval, R0, and genomic analysis, are deliberately excluded from the pilot phase due to time, budget, and size constraints. The focus is limited to achievable primary outcomes like SIR and SCAR, thus a part of scope limitation.

D. Network Expansion and Collaborative Expertise (Satellite Sites)

As a Unity Studies network site, HIMSR's responsibility extends to peripheral institutions or "Satellite Sites" to strengthen and expand the Unity Studies Network across India. This network expansion is essential for fulfilling the strategic goal of achieving wide geographical and demographic representation, a critical component of global preparedness [2].

1. Proposed Satellite Sites: The satellite sites will perform the program tasks of exercising Unity protocol, in consultation with the Nodal network site. HIMSR has proposed and invited several premier academic and research institutions across India as potential satellite sites for coordinating future HHTI exercises. These institutions demonstrated their commitment and research capacity during the COVID-19 Unity Studies and will utilize their experiences.

This distributed network strategy ensures the Nodal site can rapidly scale up surveillance efforts in diverse geographical and health system contexts when a novel pathogen emerges (Figure 3).

https://doi.org/10.3390/v18020198

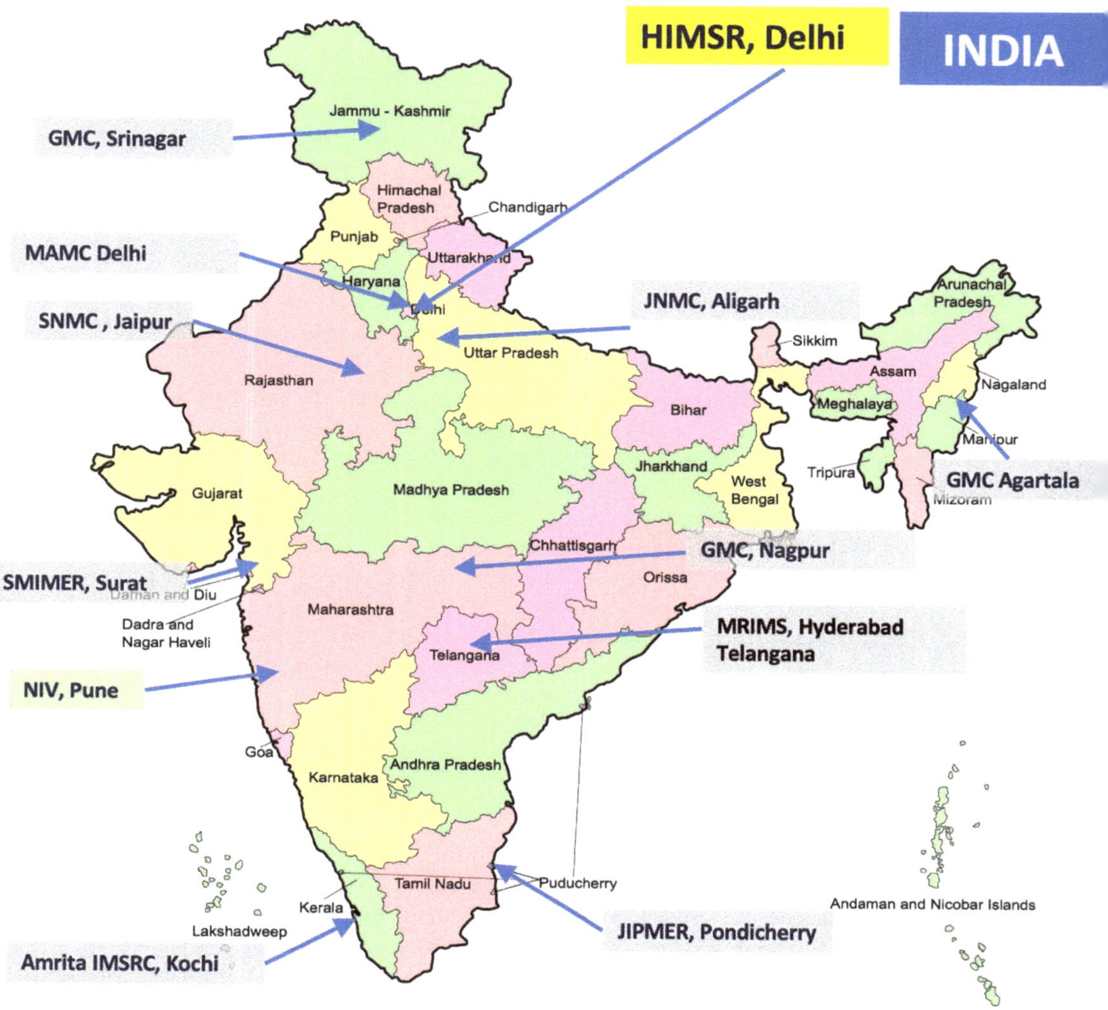

Figure 3. Medical institutes proposed for inclusion in the HHTI Unity Network, India. Footnote: Hamdard Institute of Medical Sciences and Research (HIMSR), New Delhi India; Government Medical College, Srinagar; Maulana Azad Medical College, Delhi; Amrita Institute of Medical Sciences, Kochi; Government Medical College, Agartala; Malla Reddy Institute of Medical Sciences (MRIMS), Hyderabad; SMIMER Medical College, Surat; Government Medical College, Nagpur; Jawaharlal Nehru Medical College, Aligarh; S.N. Medical College, Jaipur; Jawaharlal Nehru Institute of Post Graduate Medical Education & Research (JIPMER), Pondicherry, and ICMR- National Institute of Virology (NIV), Pune. Sites highlighted with Grey color are proposed satellite sites and with green color is WHO reference lab.

2. Nodal network site Responsibilities and Requirements: To ensure the effective function of the network, all Unity Studies Network Sites, including HIMSR, must adhere to stringent duties outlined in the Terms of Reference (ToR). These duties include the following [19]:

- Designated Focal Point: The site must have a designated site Principal Investigator (PI)/Focal Point to coordinate investigations and liaise with the WHO Country Office.

https://doi.org/10.3390/v1802

- Technical Participation: Sites must participate in relevant conference calls to identify and collaboratively solve issues early in the investigation phase, promoting international collaboration and sharing preliminary global findings.
- Research Translation: By exercising standardized protocols and establishing multi-institutional collaborations, all Unity Studies Network Sites are strengthening the research–policy–practice interface, promoting the generation of evidence that can influence policy, thereby addressing a critical need in LMICs where translation supporting structures are traditionally weak.
- Knowledge Sharing and Laboratory Links: The network aims to utilize the isolated expertise across India by creating a platform for shared learning and joint mentorship among participating sites. Another important benefit is that sites would have access to a designated a national-level laboratory which is a WHO Collaborating Center and National Influenza Centre of India and WHO Global H5 reference lab under GISRS (NIV, Pune) [19]. In addition, the sites must be willing and prepared to share isolates and specimens with the WHO Collaborating Centers.
- Data Analytics and Expertise: Sites require access to a dedicated data analytics team to perform rapid analysis. The HIMSR team's analytical capacity, supported by WHO India and WHO HQ, involving complex analysis, is foundational to fulfilling this requirement.

3. Benefits of Participation and Equity: Participation in the Unity Network provides significant benefits, particularly for LMICs [2,15]. Sites are recognized globally as a center of expertise, enabling the unlocking and mobilization of knowledge across geographic and disciplinary boundaries. The network facilitates capacity building by offering skills training in study implementation, laboratory testing, data management, and statistical/modeling analyses [2]. The overall initiative promotes research equity by ensuring that high-quality, standardized research is feasible and supported in LMICs, reducing the historical publication bias towards high-income countries. The WHO provides access to standardized and harmonized tools, such as generic field questionnaires, data collection tools, and data analysis scripts, promoting interoperability among sites [16].

E. Bridging Research and Systems Capacity

The function of the HIMSR network site extends beyond data collection to ensure the effective translation of research findings into policy. This aligns with wider national acknowledgment of the importance of building Health Systems Research.

1. Science Translation for Decision-Makers: The role of government, health authorities, and policymakers in sustained engagement within the Unity Studies Network is centered on transforming high-quality research into actionable public health policy and securing operational support. Timely dissemination of findings is crucial to update guidance and inform local, national, and international public health responses. The ultimate goal is for the standardized HHTI data to be available to decision-makers and key stakeholders to inform timely policy-relevant decisions [16]. The standardized HHTI data must be made available to decision-makers and key stakeholders to enable them to inform timely, policy-relevant decisions. This translational goal ensures that timely dissemination of findings updates guidance and informs local, national, and international public health responses. To facilitate this, the Unity protocols provide toolkits and training on how to communicate with different audiences and address policy-makers' questions.

2. Addressing Methodological Rigor for Aggregation: For the data collected by the HIMSR and its network to be pooled and effectively contribute to global policy [2,10], methodological rigor must be consistently high [6]. The WHO Unity team developed specialized tools, such as the HHTI Critical Appraisal Checklist, consisting of 12 questions covering 10 aspects like context, household definition, and follow-up quality, to assess

potential for bias and suitability for aggregation [6]. The HIMSR commits to ensuring its implementation meets these high standards.

3. Sustained Engagement: The establishment of the Regional network site contributes to overcoming the historical insularity of expertise. By fostering formal collaborations between diverse institutions, the network site promotes a necessary collective level of capacity building. This mechanism helps ensure that the generated evidence is locally relevant and improves the research–policy–practice interfaces, which are often traditionally weak in LMICs. The lessons learned from other Indian capacity-building initiatives, such as the need to align research topics with policy-makers' interests for better buy-in, are intrinsically integrated into the network site's mission to provide timely, actionable data during future outbreaks [13].

F. Budgetary and Ethics Considerations

Sustained operation requires robust financial planning. With the changing health financing global scenario and reduction in WHO financial support to the LMIC, the need for internal financial resource generation and co-funding by the network institutions for sustainable project implementation highlighted the scale of resources required. With the previous experiences during the COVID-19 HHTI study, the largest cost components were for testing samples, followed by costs for Infection Prevention and Control (IPC) materials, transport, and personnel costs. The Unity Studies network site operates on catalytic WHO support, emphasizing co-funding by HIMSR to ensure sustainability and country ownership. Furthermore, the WHO commits to providing support for protocol finalization, financial oversight, monitoring, analysis, and dissemination of data.

The successful conduct of the HHTI requires meticulous adherence to Infection Prevention and Control (IPC) measures, necessitating consistent funding for PPE and waste disposal. Adhering to ethical standards, including obtaining written informed consent in the local language, adequate data management and maintaining participant confidentiality through assigned identification numbers are also resource intensive and non-negotiable. Furthermore, the study must undergo rigorous ethical clearance processes, including approval from local (HIMSR) and WHO SEARO Ethics Review Committees for the study. Focusing on the principle of justice, the operational philosophy of the Unity Studies Network is based upon the global objective of promoting research equity.

7. Conclusions

The HIMSR team's rigorous execution of three distinct WHO Unity protocols between 2020 and 2021—HHTI, HCW cohort, and VE study—marked a significant contribution to both national and global understanding of SARS-CoV-2 dynamics, validating the critical role of standardized operational research in pandemic response. Key findings and vital epidemiological data provided immediate, context-specific evidence necessary for refining and implementing effective local public health measures.

The program effectively developed robust technical and human resource capacities, especially through close collaboration and networking with the WHO (HQ, SEARO, India Country Office) and utilization of global tools. Critical lessons learned included the necessity of managing confounding variables and addressing the methodological complexity of estimating transmission parameters in high-contact settings.

This experience provided the necessary operational capacity and intellectual foundation for HIMSR's designation as a WHO Unity Studies Network Site. The strategic transition focuses on sustaining preparedness through the 2025 HHTI pilot study on endemic respiratory pathogens (Influenza/RSV). By rigorously adhering to standardized protocols, focusing on practical capacity building, and actively expanding a national network of collaborating institutions, the HIMSR ensures that India develops a resilient

https://doi.org/10.3390/v1802019

operational research infrastructure capable of rapidly providing standardized, high-quality, and contextually relevant data to guide effective public health responses against future respiratory threats.

Author Contributions: Conceptualization, Y.A., F.I., M.A., R.G., M.H., N.L.B., A.B. and P.R.W.; methodology, Y.A., F.I., M.A., R.G., A.S., M.H., K.Y., M.R., S.B., N.L.B., A.B. and P.R.W.; investigation, Y.A. and F.I.; resources, Y.A., F.I., M.A., R.G., and M.H.; data curation, Y.A., F.I., R.G., and M.H.; writing—original draft preparation, Y.A., F.I., M.A., R.G., A.S., M.H., K.Y., M.R., S.B., N.L.B., A.B. and P.R.W.; writing—review and editing, Y.A., F.I., M.A., R.G., A.S., M.H., K.Y., M.R., S.B., N.L.B., A.B. and P.R.W.; visualization, Y.A.; supervision, F.I., A.S., and M.H.; project administration, Y.A. and F.I.; funding acquisition, F.I., M.A., M.H. and P.R.W. All authors have read and agreed to the published version of the manuscript.

Funding: This research was funded by the World Health Organization under the Pandemic Influenza Preparedness Partnership Contributions (PIP-PC) Work Plan for 2024–2025 and the APC was funded by the World Health Organization.

Institutional Review Board Statement: The study was conducted in accordance with the Declaration of Helsinki and approved by the Institutional Ethics Committee of Hamdard Institute of Medical Sciences and Research, New Delhi (HIMSR/IEC/00378/2025, dated 15 September 2025) and WHO SEARO Ethics Review Committees (2025.19.ind dated 15 September 2025).

Informed Consent Statement: Informed consent was obtained from all subjects involved in the study.

Data Availability Statement: No new data were created or analyzed in this study. Data sharing is not applicable to this article.

Acknowledgments: We are thankful to the Delhi state health department, South-East Delhi district authority, South Delhi district authority for technical support during the COVID-19 Unity study. We are thankful to the WHO HQ, SEARO, and India county office for technical and financial support.

Conflicts of Interest: The authors declare no conflicts of interest. The funders had no role in the design of the study; in the collection, analyses, or interpretation of the data; in the writing of the manuscript; or in the decision to publish the results.

Abbreviations

CI	Confidence Interval.
CONSISE	Consortium for the Standardization of Influenza Seroepidemiology.
COVID-19	Coronavirus Disease 2019.
FFX	First Few X Cases and Contacts Protocol.
GISRS	Global Influenza Surveillance and Response System
HAHC	Hakeem Abdul Hameed Centenary Hospital.
HCW	Health Care Worker or Health Worker.
HHC	Household Contacts.
HIMSR	Hamdard Institute of Medical Sciences and Research.
HHTI	Household Transmission Investigation.
ICMR	Indian Council of Medical Research.
IPC	Infection Prevention and Control.
IP	Incubation Period.
LMIC	Low- and Middle-income Country
NABL	National Accreditation Board for Testing and Calibration Laboratories

https://doi.org/10.3390/v18020198

OR	Odds Ratio.
PI	Principal Investigator.
PPE	Personal Protective Equipment.
PRET	Planned Preparedness and Resilience for Emerging Threats.
RSV	Respiratory Syncytial Virus.
RT-PCR	Reverse Transcription Polymerase Chain Reaction.
SAR	Secondary Attack Rate.
SARS-CoV-2	Severe Acute Respiratory Syndrome Coronavirus 2.
SEARO	South-East Asia Regional Office.
SIR	Secondary Infection Rate.
ToR	Terms of Reference.
VE	Vaccine Effectiveness.
WHO	World Health Organization.
WHO HQ	WHO Headquarters

References

1. Islam, F.; Alvi, Y.; Ahmad, M.; Ahmed, F.; Rahman, A.; Singh, F.H.D.; Das, A.K.; Dudeja, M.; Gupta, E.; Agarwalla, R.; et a Household transmission dynamics of COVID-19 among residents of Delhi, India: A prospective case-ascertained study. *IJID Re* **2023**, *7*, 22–30. [CrossRef] [PubMed]
2. Bergeri, I.; Boddington, N.L.; Lewis, H.C.; Subissi, L.; von Dobschuetz, S.; Rodriguez, A.; Jara, J.; El Naja, H.A.; Barakat, A Rashidian, A.; et al. WHO's Investigations and Studies, Unity Studies: A global initiative creating equitable opportunities fo enhanced surveillance, operational research, capacity building, and global knowledge sharing. *Influenza Other Respi. Viruses* **202** *18*, e13256. [CrossRef] [PubMed]
3. WHO. Household Transmission Investigation Protocol for Coronavirus Disease 2019 (COVID-19) (Version: 2.2). 2020. Avai able online: https://iris.who.int/server/api/core/bitstreams/481af62d-bb64-40fa-b836-09bfc41500de/content (accessed o 11 October 2025).
4. WHO. Protocol for Assessment of Potential Risk Factors for Coronavirus Disease 2019 (COVID-19) Among Health Workers in Health Care Setting (Version: 2.2). 2020. Available online: https://www.who.int/publications/i/item/protocol-for-assessme of-potential-risk-factors-for-2019-novel-coronavirus-(2019-ncov)-infection-among-health-care-workers-in-a-health-care-setti (accessed on 11 October 2025).
5. Bergeri, I.; Lewis, H.C.; Subissi, L.; Nardone, A.; Valenciano, M.; Cheng, B.; Glonti, K.; Williams, B.; Abejirinde, I.O.O.; Simnicear A.; et al. Early epidemiological investigations: World Health Organization Unity protocols provide a standardized and time international investigation framework during the COVID-19 pandemic. *Influenza Other Respi. Viruses* **2022**, *16*, 7–13. [CrossRe [PubMed]
6. Price, D.J.; Spirkoska, V.; Marcato, A.J.; Meagher, N.; Fielding, J.E.; Karahalios, A.; Bergeri, I.; Lewis, H.; Valenciano, M.; Pebo R.; et al. Household transmission investigation: Design, reporting and critical appraisal. *Influenza Other Respi. Viruses* **20** *17*, e13165. [CrossRef] [PubMed]
7. Dudeja, M.; Shaikh, A.; Islam, F.; Alvi, Y.; Ahmad, M.; Kashyap, V.; Singh, V.; Rahman, A.; Panda, M.; Shree, N.; et al. Assessme of Potential Risk Factors for COVID-19 among Health Care Workers in a Health Care Setting in Delhi, India—A Cohort Stu *PLoS ONE* **2023**, *18*, e0265290. [CrossRef] [PubMed]
8. Das, A.K.; Islam, F.; Alvi, Y.; Dudeja, M.; Ahmad, M.; Rahman, A.; Roy, S.; Ahmed, M. SARS-CoV-2 infection and seropositiv among household contacts of laboratory confirmed cases of COVID-19 in residents of Delhi, India. *Prev. Med. Rep.* **20** *38*, 102603. [CrossRef] [PubMed]
9. Bhatnagar, T.; Chaudhuri, S.; Ponnaiah, M.; Yadav, P.D.; Sabarinathan, R.; Sahay, R.R.; Ahmed, F.; Aswathy, S.; Bhardwaj, Bilimale, A.; et al. Effectiveness of BBV152/Covaxin and AZD1222/Covishield vaccines against severe COVID-19 and B. 1.6 2/Delta variant in India, 2021: A multi-centric hospital-based case-control study. *Int. J. Infect. Dis.* **2022**, *122*, 693–702. [CrossR [PubMed]
10. Lewis, H.C.; Marcato, A.J.; Meagher, N.; Valenciano, M.; Villanueva-Cabezas, J.P.; Spirkoska, V.; Fielding, J.E.; Karahalios, Subissi, L.; Nardone, A.; et al. Transmission of SARS-CoV-2 in standardised First Few X cases and household transmiss investigations: A systematic review and meta-analysis. *Influenza Other Respir. Viruses* **2022**, *16*, 803–819. [CrossRef] [PubMed
11. Dudeja, M.; Sharma, P.; Islam, F.; Shaikh, A.; Singh, F.H.; Alvi, Y.; Kashyap, V.; Mariam, W.; Das, A.K.; Haque, S.F.; et al hospital-based prospective cohort study to assess the factors associated with transmission dynamics of SARS CoV-2 amo healthcare workers in Delhi. *F1000Research* **2022**, *11*, 728. [CrossRef]

https://doi.org/10.3390/v1802

12. WHO. Strengthening Pandemic Preparedness Through Unity Studies: Real-World Applications from India in WHO's South-East Asia Region. 14 October 2025 Departmental Update New Delhi, India. 2025. Available online: https://www.who.int/southeastasia/news/detail/14-10-2025-strgthng-pandemic-Unity-studies (accessed on 20 October 2025).
13. WHO. Seventy-Eighth World Health Assembly Agenda Item 16.2: WHO Pandemic Agreement. World Health Organization: Geneva, Switzerland, 2025. Available online: https://apps.who.int/gb/ebwha/pdf_files/WHA78/A78_R1-en.pdf (accessed on 16 October 2025).
14. WHO. Mosaic Respiratory Surveillance Framework. Surveillance for Respiratory Viruses of Epidemic and Pandemic Potential. World Health Organization. 2023. Available online: https://www.who.int/initiatives/mosaic-respiratory-surveillance-framework (accessed on 2 October 2025).
15. Unity Studies Global Network: A Global Initiative to Strengthen Pandemic Preparedness Response. P.A.H.O./W.H.O. 2025. Available online: https://www.paho.org/en/news/6-10-2025-Unity-studies-global-network-global-initiative-strengthen-pandemic-preparedness-and (accessed on 20 October 2025).
16. Unity HHTI Protocol. Household Transmission Investigation (HHTI) Template Protocol for Respiratory Pathogens with Pandemic Potential. World Health Organization. 2023. Available online: https://cdn.who.int/media/docs/default-source/documents/epp/respiratory-unity-study/unity-studies_g2_hhti.pdf?sfvrsn=fe7b1adc_2 (accessed on 1 October 2025).
17. Islam, F.; Alvi, Y.; Gautam, R.; Zenodo. Household Transmission Investigation (HHTI) & Associated Risk Factors for Influenza A, Influenza B and Respiratory Syncytial Virus (RSV) Infection Amongst Residents of Delhi, India. 2025. Available online: https://zenodo.org/records/17446050 (accessed on 2 September 2025).
18. Dutta, A.; Jinsart, W. Air pollution in Delhi, India: It's status and association with respiratory diseases. *PLoS ONE* **2022**, *17*, e0274444. [CrossRef] [PubMed]
19. WHO. Terms of Reference for Unity Studies Network Sites in the WHO SEARO Region (September 2024). 2024. Available online: https://cdn.who.int/media/docs/default-source/documents/epp/respiratory-unity-study/tor_unity-studies-2.0-network-sites_sept-2024ea2f2186-44a2-4354-bdbc-c25e72b376b4.pdf?sfvrsn=b4c7963b_3 (accessed on 2 September 2025).

https://doi.org/10.3390/v18020198

MDPI AG
Grosspeteranlage 5
4052 Basel
Switzerland
Tel.: +41 61 683 77 34

Viruses Editorial Office
E-mail: viruses@mdpi.com
www.mdpi.com/journal/viruses

www.ingramcontent.com/pod-product-compliance
Lightning Source LLC
Chambersburg PA
CBHW062036220926
51850CB00010B/2572
9783725878338